FUTURE WAR

FUTURE WAR

Preparing for the New Global Battlefield

Robert H. Latiff

ALFRED A. KNOPF | NEW YORK | 2017

THIS IS A BORZOI BOOK
PUBLISHED BY ALFRED A. KNOPF

 Published in the United States
by Alfred A. Knopf, a division of Penguin Random House LLC,
New York, and distributed in Canada
by Random House of Canada, a division
of Penguin Random House Canada Limited, Toronto.

www.aaknopf.com

Library of Congress Cataloging-in-Publication Data

Names: Latiff, Robert H., author.
Title: Future war : preparing for the new global battlefield /
by Robert H. Latiff.
Description: Preparing for the new global battlefield | First edition. |
New York : Alfred A. Knopf, 2017. | "A Borzoi book." |
Includes bibliographical references and index.
Identifiers: LCCN 2017002411 |
ISBN 9781101947609 (hardcover) |
ISBN 9781101947616 (ebook)
Subjects: LCSH: Military art and science—United States—
History—21st century. | Military art and science—
Moral and ethical aspects. | War—Moral and ethical aspects. |
War—Forecasting.
Classification: LCC U43.U4 L37 2017 | DDC 355.0201/12—dc23
LC record available at https://lccn.loc.gov/2017002411

Jacket image: chuyuss/Shutterstock
Jacket design by John Vorhees

Manufactured in the United States of America

First Edition

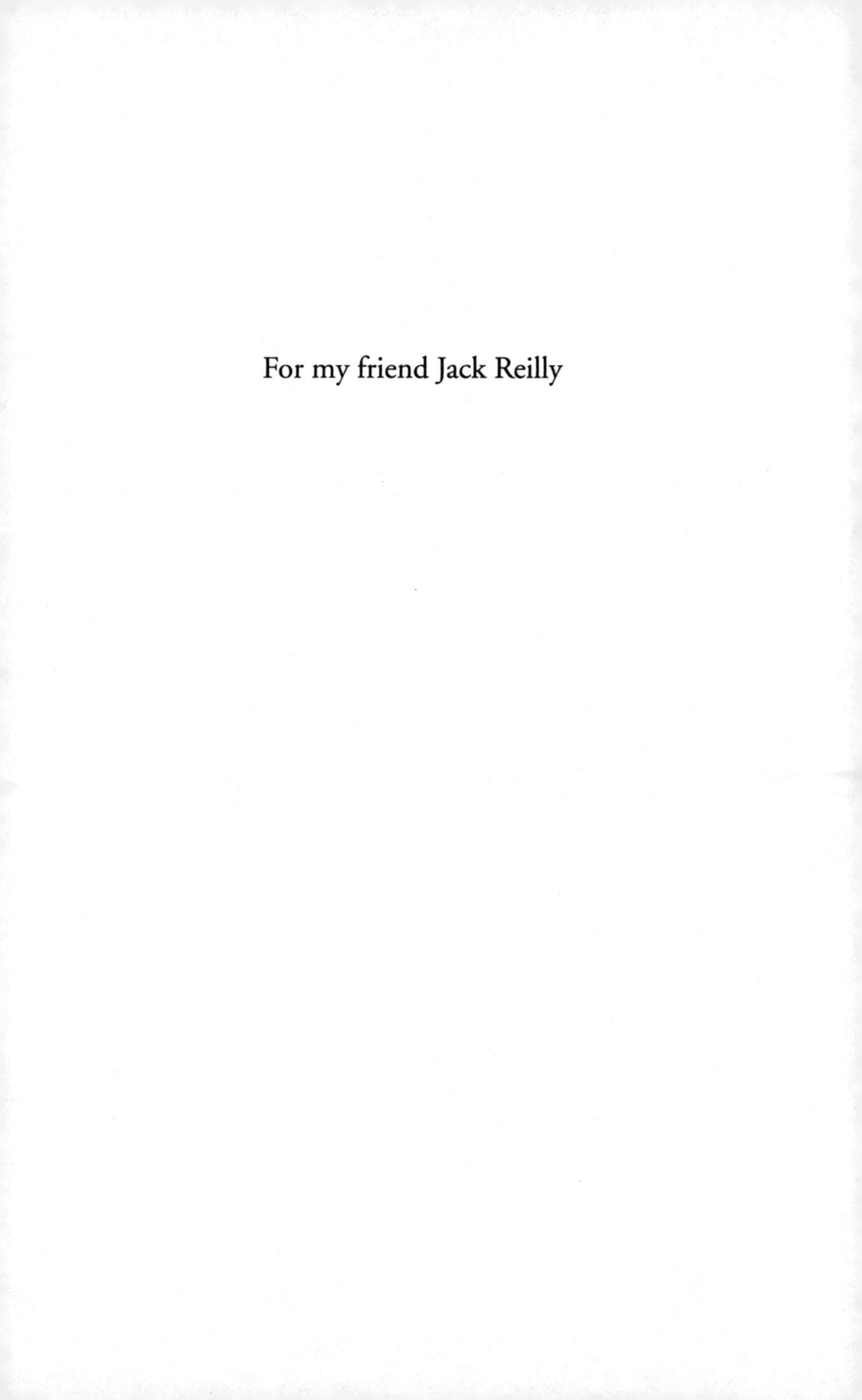

For my friend Jack Reilly

If we continue to develop our technology without wisdom or prudence, our servant may prove to be our executioner.

Ours is a world of nuclear giants and ethical infants.

—General of the Army Omar N. Bradley,
Armistice Day speech, November 1948

CONTENTS

FUTURE WAR

INTRODUCTION

ON A SWELTERING AUGUST DAY, *following weeks of heightened tensions with Russia over its actions in Ukraine and Syria, and harsh words between the United States and China over actions in the South China Sea, and as workers are preparing to depart the cool of their air-conditioned Manhattan office buildings for the gridlocked highways and subways, several large electric power plants along the East Coast simultaneously experience dramatic over-speed conditions in their large turbine motors. Plant operators are unable to stop the steam turbines, whose automatic control and data systems have been infected by sophisticated computer malware, and they catastrophically tear themselves apart, cutting power to large segments of the population and industries in the Northeast. Building systems shut down, hospitals switch to emergency generators, trains stop running, traffic lights cease operation, and Wall Street trading comes to a halt. At the same time, eleven hundred miles to the south, a massive rocket sits fueled and ready to launch a critical national security satellite, when an explosive-laden private aircraft flies at low altitude into the Cape Canaveral launch area and, despite repeated warnings,* *slams into the pressurized fuel tanks and solid rocket motors, creating an enormous conflagration. Half a world away, elite commandos equipped with the latest high-technology equipment but unidentifiable as a national army attack U.S. and allied interests near areas of disputed territory. Thus are fired the opening shots of a new war.*

These hypothetical events represent a radically different style of conflict, fought with new tools and against new and unfamiliar enemies. When most people think of war, they imagine soldiers doing battle with other soldiers, employing tanks, artillery, and other recognizable weapons. However, in this century war has morphed into something we can scarcely recognize, and future conflicts will be qualitatively and quantitatively unlike those of the past. They will be fought using innovative and unusual weapons, many of which, because the technologies have both civilian and military uses, will be available to far more people who are far less skilled in their employment. The so-called democratization of technology has diminished the monopoly of advanced countries on the tools of war.

Twenty-first-century armed conflicts often have no battlefield in the traditional sense. The concept of opposing armies clashing in deadly struggle but moderated by international conventions of behavior seems a thing of the past. In 1999, Chinese colonels Qiao Liang and Wang Xiangsui predicted that soldiers would increasingly be computer hackers, financiers, drug smugglers, and agents of private corporations rather than members of a military, and that their weapons would include not only airplanes, cannons, poison gas, bombs, and biochemical agents but also computer viruses, net browsers, and financial derivative tools. Their predictions were prescient.

Yesterday's wars were, like World War II, about saving civilized nations from maniacal dictators, or, like the conflicts in Vietnam and Korea, devoted to the clash of ideologies and the attempt to limit the spread of one hegemon over another. They were big affairs, involving large military forces and enormous violence. Today's wars are more about cultural and religious hatreds, using violence as a means to change the hearts and minds of people, among whom the killing occurs with more frequency. Tomorrow's wars will be different still, fought largely for political dominance with stealth and cunning, targeting innocents and institutions, heavily dependent on information superiority, and employing strange new weapons.

While we have highly motivated and well-trained and -equipped armed forces that will adapt, we are not prepared as a nation to react well to the inevitably messy and ambiguous situations these new conflicts and weapons will present. We still have no clear idea of what constitutes an act of war in cyberspace nor of how we might respond to an attack. It is not enough for the military to be prepared. Neither the public nor its decision makers have yet fully comprehended the significance of the changes in the types of conflict and the tools with which they will be fought.

Without question there will be circumstances when decision makers will find it necessary, or at least think it necessary, to send soldiers into classic "boots on the ground" combat situations. We will continue to project force with our powerful aircraft, our munitions, and our aircraft carrier strike groups.

When war is fought in foreign lands, it could well be against other high-tech enemies. However, war will increasingly be more personal and often fought closer to home. It will not lend itself to the traditional massive displays of U.S. firepower. It will affect individuals directly, not as some distant conflict we read about in newspapers or watch in the movies or on the Internet.

Americans will be targeted on U.S. soil, often, as we have seen in San Bernardino and Orlando, by homegrown terrorists. This form of war will be messy and complex, and it will not lend itself to quick, easy, sound-bite solutions. It may not even be clear for a while who perpetrated the violence against us. Was it state-sponsored, or a random act of terror? Will we unleash massive destruction on foreign countries in retaliation for attacks by terrorists residing on our own soil? If so, to what end? And with what consequences? There are always consequences, whether they be the death or capture of our own troops, unnecessary civilian casualties, or the incitement of more violence. Whether we respond in thoughtless, knee-jerk fashion, lashing out at the first available target, or with professional military efficiency will be determined by whether or not our leaders have the will and the intellectual capacity to think logically rather than act rashly. Too often we have been dragged by our political leaders into conflicts—in Libya for instance—with too little analysis of the future ramifications.

Technology and war are topics of critical importance to everyone today. The conflicts in which we are currently engaged are enabled by new technologies such as the Internet, social media, and rapid worldwide communications. We saw this on display in the massive use of social media during the so-called Arab Spring uprising in Egypt. More recently, Russia has used computer attacks and social media in its annexation of the Crimea and its support of rebels in Ukraine. A large part of the world's daily life is determined by technology of some sort. By any measure, the last century and a half has seen an explosion of technological innovation and development, much but not all of which has bettered the human condition. Sadly, the period has also seen monstrously destructive wars and a seemingly never-ending proliferation of arms and violence around the world. These two developments—technological advancements and destructiveness in war—are now more codependent than ever before.

Technology and war have become far more complicated, and understanding them will require hard work. Wars of the future will be full of complexity and ambiguity. Weapons will employ such new technologies as artificial intelligence and synthetic biology, neither of which we yet fully understand. The effects of technological advancement in these two areas are far more difficult to predict, and possibly to control, than are conventional weapons. Other weapons, including lasers and radio-frequency technologies, are not only instantaneous in their effects, but the damage they may cause is incompletely understood. These new technologies are changing in fundamental ways how we fight.

Killing has gone from military unit against military unit to the targeting of individuals. It has gone from the control of land and large standing armies to the control of computers and networks. It may go from the use of massive chemical-based explosives to the use of destructive microbes.

The new technologies of war are tremendous assets to the military. They provide soldiers with advantages over the adversary and also greatly reduce the risks soldiers face. They also present ethical challenges in their application. There are strong rationales and value for these weapons, but also some obvious and some not so obvious disadvantages and ethical concerns with how they will challenge the individual soldier's behavior in war. With the deployment of any new technology, whether in the civilian or military worlds, come uncertainties about its use, effects, and attendant risks. The unchecked and inadequately considered development, adoption, and proliferation of high-tech weapons will surely have unintended and undesirable consequences that we cannot know in advance, but that could include environmental damage, risks to the soldier himself, or a more dangerous response by adversaries.

Technology and weapons have had a long and symbiotic history. The weapons that are proliferating in the world today, and therefore the people who will control them, are more dangerous than any we've yet encountered. Once unleashed, they may be uncontrollable. While in the Cold War the superpowers acted in somewhat predictable, rational ways, we consider current adversaries like the Islamic State or North Korea to be "irrational" actors who behave in unpredictable and excessively brutal ways. Potential adversaries who do not currently have advanced technologies such as nuclear weapons and biological agents will seek them. While we deal with emerging adversaries, our old, technologically capable foes are engaging us in a deadly arms race for advantage in both current and future technologies.

The technological superiority we so highly prize and cultivate, while crucial to our success, breeds in us a dangerous arrogance. Technology and the weapons it spawns seduce us, addict us, and often invite hubris. We frequently overestimate the advantages a new technology will provide. There is also a danger that if we fail to fully question the implications of our new weapons, if we fail to think about the long-term effects they may engender, we may unleash something we will wish we hadn't. We had better understand these complex new weapons, the military, and our adversaries before we again bully our way into another war, thinking our technology will save the day.

New types of conflict and the rapid growth of technology will challenge both soldiers and the public in their understanding and their responses. Reasons to go to war, and behavior in war, are important to civilized nations. Militaries and the public they serve should be sensitive to the laws of armed conflict and the resulting body of international humanitarian law that embodies them, lest they undo centuries of civilized behavior. The incorporation of new methods and tools of war will certainly result in dilemmas as we attempt to adapt to them. New technologies, like autonomous systems—those that perform their functions without any human control or intervention—chip away at our humanity and must be more carefully considered before being employed in weaponry.

How should military leaders, scholars, legislators, and the public react to the uncertainty that comes with technological advance and the new kind of war it creates? Have new weapons technologies and types of conflict made obsolete the traditional frameworks for thinking about the ethics of war and peace? Will leaders understand well enough the parameters of an issue and the relevant criteria for making a decision? These questions may seem cumbersome, but the idea of international norms and ethics forms the basis for the military services' core values and modern concepts of the warrior ethos. There are important norms of behavior and commonly accepted rules of war that remain important and relevant, even in the face of breathtaking technological change and the unspeakable brutality of our adversaries.

The public can no longer stand on the sidelines while a small percentage of the population does the dirty work of war. Sadly, few Americans have a proper understanding of war and technology, and are even less aware of or interested in the ethics of warfighting. The U.S. military is increasingly a thing apart from the American public. The people and their leaders display a stunning lack of interest and knowledge. They "support the troops" but know precious little about them or their missions. This deliberate ignorance is not only shameful, it is dangerous. In the absence of clear and unambiguous public involvement, the military will respond to events in the way it deems most appropriate, and it is unconscionable for a nation to cede its most important decisions to such a small and highly homogeneous group. For this reason alone, the American public must end its dangerous complacency about issues of war. There is a large and growing chasm between the military and society in general, and a skewed view of the military by the public, partly a result of a romanticized, entertainment-fueled notion of war and soldiering and an outsized view of the capabilities wrought by technology. This chasm must be bridged.

Our leaders, who themselves for the most part have only a cursory knowledge of technology or military affairs, continue to deploy our forces with little restraint. In the future, the American people and their leaders must be intimately and actively involved in deciding how and when those forces act, and with what weapons. If they don't start paying more attention, we will again be at risk of self-delusion and costly decisions about the use of force. Everyone will be affected by future conflicts fought with new technology. Every citizen should understand and care about why we risk the lives of our young men and women, with what tools we will fight future enemies, and how our soldiers behave on the battlefield. One of my main reasons for writing this book is to encourage more active involvement and informed and thoughtful analysis by national leaders, and the citizens they serve, before soldiers, sailors, airmen, and Marines are committed to combat.

If someone asked me fifty years ago if I was concerned about the ethics of warfighting and weapons technologies or the behavior of our military, my answer would likely have sounded very much like that of anyone from my conservative southern Kentucky hometown. "Our country, right or wrong." "America, love it or leave it." I was a product of a family of war veterans and came of age during the Sputnik generation. Without a U.S. Army ROTC scholarship, I would have had to drop out of college, which would have cost me not only a degree in physics, but also a keen sense of social justice and ethical awareness, along with a strong dose of questioning and healthy cynicism.

Entering the Vietnam-weary Army after earning a Ph.D. in engineering, I served a tour in the infantry in Germany, guarding the Fulda Gap, through which hordes of Soviet soldiers were reportedly planning to overrun western Europe. I then commanded a weapons unit that held the "tactical" nuclear warheads we would use to stop the Soviets if our conventional forces failed. It was then that I began to wonder about the madness of what we were doing and might be asked to do. Looking back on some of the provocative things both sides did, like deliberately flying close to or over the border, I am amazed we never came to blows. Being a good soldier, I would surely have carried out whatever my orders may have been, but it rattled me.

After a transfer to the U.S. Air Force, where I could use my relatively new doctorate, I quickly became enmeshed in research on unusual new materials that I would later learn were intended for a new generation of stealthy aircraft. I ultimately ended up working at the Pentagon on President Reagan's Strategic Defense Initiative ("Star Wars") program. With more money than we knew what to do with, we started all sorts of interesting and scary programs, such as nuclear-pumped space-based lasers. In theory, a nuclear weapon detonated in space would energize multiple lasers to shoot down enemy ballistic missiles. It was all exciting and career-enhancing, though it didn't take much to realize that it wasn't really going to work. Ironically, the idea of it alone was very frightening to the Soviets, and therefore highly destabilizing and effective.

During the year in which America fought the Gulf War and the Berlin Wall fell, I was attending Harvard's John F. Kennedy School of Government, where I studied history, strategy, and international economics. My cynicism increased. I began to understand far more clearly the enormous impact of personalities, politics, and money on the deadly serious business of war.

After a series of assignments directing larger and larger intelligence, surveillance, and reconnaissance programs, ground-based telescopes and radars, and big radar airplanes, I was promoted to brigadier general and sent to command one of the most iconic of Cold War facilities, the underground command center at Cheyenne Mountain, Colorado. There, we literally watched over the world—and space—and reported unusual happenings to the national command authorities. We also practiced daily for what we would do in case of nuclear attack. I was way, way down on the order-of-precedence list, but at some point after a nuclear exchange I *might* have had to recommend a response to the president. It was a troubling responsibility. I left not long before September 11, 2001. My friend and next-door neighbor, a recently retired Air Force colonel, was killed on the first airplane to hit the World Trade Center. In my position as vice commander of a large Air Force center, I signed stop-loss orders forcing airmen to remain on active duty past their original separation dates, and sent airmen to Afghanistan.

My last assignment was to the National Reconnaissance Office (NRO), where the nation's spy satellites are designed, built, and operated. I directed the development of very advanced surveillance technologies of all types for watching and listening to our adversaries, and worked on joint projects with similar technology offices at CIA, NSA, and other intelligence agencies, as well as the Department of Defense (DOD). Everyone was correctly focused on the mission of hunting down and eliminating the 9/11 terrorists and others like them. On occasion, however, I felt that in the pursuit of new methods the United States was far too aggressive. We were often dismissive of privacy and civil liberties concerns, pushed the boundaries of ethics in technologies used to track individuals, and pursued techniques that we would never want others to use against us. While at the NRO, I was also the senior officer and commander for the large number of Air Force personnel there. As such, I repeatedly deployed people to the Middle East, including Iraq.

Since retirement from active military service in 2006, I have consulted with military and intelligence agencies on a broad range of technologies. I followed the Iraq War with dismay as it spun out of control, the insurgency fueled by the revelations of torture and kidnapping by the United States. I cringed at the reports of warrantless wiretapping and other abuses of the government's surveillance powers. I was embarrassed by U.S dismissiveness of our allies. After leaving the Air Force, I began to write, speak publicly, and teach courses, both at universities and to intelligence agencies themselves, on the ethics of the new styles and tools of war. This book derives from that work.

In President Obama's Nobel Peace Prize speech, he cited the importance of Just War Theory and the principles of last resort, proportionality, and discrimination. The deputy secretary of defense, the vice chairman of the Joint Chiefs of Staff, and numerous combat leaders have all also publicly discussed the importance of ethics in our conduct of future war. It is important that senior leaders accept these concepts and talk openly about them to our soldiers. More important, however, is that they know how to put them into practice.

My purpose here is to explore how new technology has changed and will change warfare. It is to highlight both the dramatic developments in technology and war and the speed with which they have occurred and to describe how these will challenge soldiers, decision makers, and the public.

As ever, some talk in terms of fear, apocalyptic scenarios, and dystopian futures. Others see the unrelenting advance of technology as the natural order of things and paint a rosy picture for mankind. I believe neither to be the case and that the truth must lie somewhere in the middle. In any event, momentous changes are coming our way in the future of warfare, and as a nation we are woefully unprepared to deal with them. Few understand what the future portends, and, scarily, few seem to care.

Fortunately, a small but growing number of scholars and government officials have begun to recognize that new technologies and new forms of warfare bring unknown dangers, and are asking how we should deal with them. I hope this book will help guide the search for answers and serve as a wake-up call for military planners, weapons technologists, decision makers, and the public as we prepare for a very different future. A new and frighteningly complex world of conflict and technology and the inevitable deadly dilemmas we will face in twenty-first-century wars demand that we pay more attention to the issues that will confront us, before it is too late to control them.

1

THE NEW FACE OF WAR

AS THE NATION *begins to absorb and attempt to adjust to the severity of the attacks on the Northeast's power grid, our satellite intelligence capabilities, and U.S. allies overseas, several key Defense Department systems come under withering cyberattack and the unthinkable happens—our nuclear command and control system goes dark. Imagery and signals intelligence satellites and aircraft come under repeated laser and high-power radio-frequency attacks, degrading or destroying sensitive electronics. Ground, naval, and air forces operating near the borders of possible adversary countries are subjected to intense electronic warfare attacks carried out by autonomous unmanned aircraft, ships, and ground vehicles operating in swarms. Distracted by multiple losses of our eyes, ears, and ability to communicate, we are oblivious to the infiltration of clandestine operators into major U.S. cities, placed there for the purpose of recruiting allies and fomenting unrest.*

Just when we thought we had reached the point of invincibility, with our highly trained military and our arsenal of expensive high-tech weaponry, along came a group of fanatics who in a single event destroyed the myth we were telling ourselves. On September 11, 2001, I was the deputy commander of one of the Air Force's weapon system development centers. I sat in my conference room at Hanscom Air Force Base near Boston with a group of scientists from one of our national laboratories, discussing collaboration on some high-tech, network-centric way of defeating advanced air defense systems. We watched the first World Trade Center tower burn and saw the second airplane hit. The irony in the simplicity of the attacks was not lost on any of us. Everyone realized on that day that things were going to change, and change they did.

This audacious attack, as well as previous attacks on U.S. embassies in Tanzania and Kenya and on the USS *Cole,* made it clear that going forward, war was going to be different. Individuals, groups, and failed states, in addition to state sponsors of such troublemakers, now threaten U.S. and allied interests and citizens globally.

The intervening years have seen massive increases in research and development and a host of new technologies for war. Since 9/11, the military has experienced the largest sustained increase in funding in its history, exceeding even that of the Reagan defense buildup. The same period also saw breathtaking advances in computers, social media, and biology, with the introduction of smartphones, Facebook and Twitter, and the decoding of the human genome.

Many of the technologies developed in recent years are dual-use, meaning they are also important in the civilian world. They are also cheaper and more readily available to our adversaries. Bombs, bullets, missiles, armored vehicles, planes, and ships have proliferated in large numbers, but weapons based on the new technologies are becoming increasingly important. Many have never before been used in combat. Unlike classic warfighting technologies like explosives, ballistics, and aerospace, they are complex and not easily understandable. They also present our soldiers and decision makers with new questions about the appropriateness of their use.

NEW TYPES OF CONFLICT

War seems to be forever changing in its form and function, from terrorists using shoe bombs and exploding underwear to research agencies developing laser weapons and providing soldiers with superhuman capabilities. New and different technologies are now finding their way into conflict, and conflict itself is changing yet again. With the change in the type and tactics of a new and different enemy, we have evolved in the direction of total surveillance, unmanned warfare, standoff weapons, surgical strikes, cyber operations, and clandestine operations by elite forces whose battlefield is global. Our attention also now turns to employing autonomous systems, enhanced soldiers, lasers, and high-speed munitions to defeat technologically sophisticated adversaries. Military and intelligence officials admit to serious concerns about nuclear terrorism and biological warfare. We must consider the reality that peer or near-peer nations will also employ high-tech systems against us in any conflict. Quite simply, the United States will need to be prepared for everything.

Recent warfare has tended toward more limited engagements in which technologies like drones and advanced surveillance have often played a critical role. Enemies will increasingly find new and clever ways to attack us. Suicide bombs and civilian airliners are ever-present threats, but information warfare and potentially lethal genetically modified viruses present new possibilities. Emerging nations and movements, knowing they cannot compete with technically advanced nations, resort to nontraditional and innovative methods and weapons. With greater access to advanced technologies, and unconstrained by rules of civilized behavior, these adversaries pose a serious threat to U.S. and allied interests, changing the balance of power in favor of what National Defense University's T. X. Hammes calls the "many and simple" against the "few and complex."

Battles of the future will not necessarily be fought on battlefields as we know them, but in cities, in ungoverned areas, in cyberspace, and in the realm of the electromagnetic spectrum. Even outer space will be a contested environment.

Terrorism and attacks on soft targets, long an issue in other countries, will become more common in America, as will the accompanying demand for better capability to counter such violence. Recent gruesome terror attacks in Paris, Brussels, San Bernardino, Orlando, Nice, Berlin, London, and Stockholm add to terrorism's already long history in third world countries and represent the future of conflict with the West. After each of these horrific incidents there were predictable calls for more, better, and increasingly intrusive intelligence gathering and for vengeance against the perpetrators.

As the opening scenario suggested, our adversaries may not always be our technological inferiors. If the rise of the Islamic

State and the continued influence of al-Qaeda and the Taliban weren't enough, old rivalries with Russia are reemerging, with that country pouring enormous sums of money into military modernization, including of its nuclear arsenal. China continues its dizzying technological advance, featuring an aggressive space program accompanied by rapid militarization. North Korea relentlessly pursues its nuclear weapons and ICBM ambitions. The United States is faced with the need to accelerate its development of new technologies and weapons to counter these emerging threats, be they real or perceived.

THE FUTURE BATTLEFIELD

In 2014, the U.S. Army Research Laboratory asked a group of experts to describe the battlefield of 2050. They predicted that it would be characterized by the presence of augmented and enhanced humans; ubiquitous robots operating in swarms and in teams with humans; automated decision making and autonomous processes, whether for weapons or for the institutional systems used to command and control them; large-scale self-organization and collective decision making by entities on the battlefield; modeling and simulation of opponent behavior; a highly contested information environment with spoofing, hacking, misinformation, and dense electronic warfare; laser and microwave weapons; and the targeting of specific individuals based upon their unique electronic and behavioral signatures. This last technology is getting closer to reality. The Defense Department is investing in technologies that can identify a specific person, at a distance, in numerous ways—

through the shape of the ear, the individual's gait, fingerprint scanners that work at a distance, chemical markers in sweat, and the specifics of a heartbeat, among others.

Soldiers will look and be different. Technology will be employed, first externally, to give the soldier greater protection, greater situational awareness, and greater stamina. Some military pilots are already being given legally approved stimulants to increase their alertness during lengthy air missions. Function-enhancing drugs will become more common. Soldiers' bodies will be modified for greater efficiency. They are likely to be artificially enhanced with exoskeletons to improve strength, drugs to improve cognition or alter memory, and surgery to implant microelectronic neurological aids. Today's deployed soldier has the capability to reach back electronically from wherever he is into vast databases of imagery and threat information as well as personal information on potential terrorists. Future soldiers will be part of a global network, tied into computers far distant, that will monitor their physical health and state of mind, determine their weapon and ammunition status, calculate their performance contribution to the unit, and issue commands.

Machines will fight for us, and at ever-greater distances. Long range, over-the-horizon, and beyond-visual-range missiles and torpedoes will be common. The commander of the Navy's submarine forces recently challenged scientists to deliver systems with ranges of two hundred miles. Instead of a torpedo being launched at a specific target, it will be sent to where a target might be in the future. Such systems are able to monitor possible target areas and respond to threats more quickly, greatly expanding the reach of friendly forces. Unmanned systems will grow in popularity and will extend their ranges. The Defense Department's inventory of unmanned aircraft, for instance, increased fortyfold in the ten-year period after 2002. With a new military emphasis on swarming drones, the rate of growth is likely to increase even further. Unmanned systems will be both controlled and autonomous, and soldiers will be able to stand off from and engage an adversary at safe distances or replenish ammunition from mobile stores.

Machines will watch for us. Almost all public spaces are now monitored by cameras, and vast amounts of digital data are now collected by both government and industry. This will only increase. Video surveillance was a $2 billion industry in 2002. It reached $21 billion in 2015. Add to the ubiquitous cameras other emerging technologies like automated identification software, face and eye scans, and radio-frequency identification tags, and the amount of data on individuals will be enormous. Sensors will be everywhere. Surveillance capabilities will depend even more heavily on drones, satellites, computer spying and exploitation, and communications intercepts, among other tools, and less on human assets.

Machines will think for us. War will be increasingly fought with computers, whether in the form of cyber war, such as an attack on infrastructure, or with computers providing support to soldiers and weapons in the field. The extensive data from sensors will be subjected to intensive mining, and artificial intelligence systems will supplant the intelligence analyst. Rather than analyzing information, the soldier will merely choose among machine-provided products. Computers will analyze streams of data collected by the soldier's clothing, equipment, and biometric sensors. Analyzed, simulated, modeled, and validated courses of action will be presented to the soldier, along with a menu of high-tech weapons. Command and control will be increasingly tied to and dependent upon a central authority.

We will fight fast. Hypersonic vehicles and munitions will dramatically shorten timelines, as will increased employment of lasers and radio-frequency weapons that operate literally at the speed of light and the lethality of which can be selected at will. Soldiers will operate under intense time pressures while deluged with information, some of it conflicting. Time will be precious. Lethal decisions will come fast. The sheer speed of battle will stress decision making.

Battles will not be well defined temporally. They will be spread out in space and time, and the adversary will rarely be readily recognizable. Battles may be subtle and take place over long periods, or they may be instantaneous and devastating. They will occur imperceptibly or at unimaginable speeds. War will not necessarily be fought between nation-states, but will be a hybrid, with individuals, groups, and nations employing various forms of violence against other individuals, groups, and nations. The Department of Defense increasingly refers to the challenges of conflict in the so-called gray zone, characterized by information operations, cyberattacks, terror, conventional attacks, and other varied forms of troublemaking. Recent experiences with Hezbollah, al-Qaeda, the Taliban, and ISIS come to mind, but we will also have to deal with organized crime groups and quasi-state-sponsored hackers.

We will fight over a larger, more diffuse battlefield, in small units of highly specialized soldiers. In some cases, the soldiers may never have to leave their base to unleash destruction on an enemy. Engagements will be customizable, with forces deployed as either individual soldiers or units. Units may consist of both normal and enhanced humans and robots.

The sphere of battle will be global and extraterrestrial, and our traditional understanding of the rules of war will be challenged. Concepts like discrimination between combatants and noncombatants, weapon proportionality, and military necessity will take on new and as yet unknown meanings.

FUTURE WARFIGHTING TECHNOLOGIES

For the future, the Department of Defense lists among its basic science research interests synthetic biology, quantum information science, cognitive neuroscience, human behavior modeling, and novel engineered materials. The Air Force lists hypersonic vehicles, laser weapons, and autonomous systems as its research priorities. In addition, the DOD plans to demonstrate new technologies and techniques in the electromagnetic spectrum, such as radio-frequency agility, and smart radios and smart antennas, which can adapt themselves to a particular need. The future battlefield will be teeming with electromagnetic impulses, some for communications, some for navigation, and some that may be harmful to humans.

Technological dominance remains the central feature of U.S. military superiority. The DOD refers to the various eras in which U.S. technology held sway over our adversaries as "offsets." The First Offset included the technology of nuclear weapons, intercontinental ballistic missiles (ICBMs), and spy satellites. The Second Offset included the introduction of stealth technology and precision-guided munitions. And now the U.S. military leadership has begun to talk openly about the Third Offset strategy. As the columnist David Ignatius summarizes, "The U.S. has obviously concluded, not surprisingly, that the best strategy is to leverage its biggest advantage, which is technology. The concepts are reminiscent of President Reagan's 'Star Wars' initiative, but thirty years on." The deputy secretary of defense describes five elements of the strategy: (1) autonomous learning systems, or machines that can adapt to changing circumstances and learn over time; (2) human/machine collaborations, in which machines help humans process vast amounts of information; (3) assisted human operations, such as wearable lightweight sensors and communications gear for soldiers on the battlefield; (4) human/machine combat teaming, in which a manned weapon partners with unmanned weapons that provide information, communications, or extra munitions; and (5) "network enabled semi-autonomous technology," which allows weapons to communicate with one another to find targets if communications or sensor links to human decision makers are destroyed. The Third Offset will depend heavily on advances in artificial intelligence and advanced computing. Obviously, the list of technologies is not complete or all-inclusive, with a great deal of research remaining highly classified.

One of the most important players in developing technologies for the Third Offset, indeed for the Defense Department in general, is the Defense Advanced Research Projects Agency (DARPA), created by President Eisenhower in 1958 to ensure that the United States is never again surprised technologically as it was by the Soviet Union's launch of the Sputnik satellite. DARPA's innovations include the stealth technology used in modern aircraft to reduce radar signatures, and the ARPANET, forerunner of today's Internet. Researchers there are often referred to by the press as the military's "mad scientists," a moniker at which director Dr. Arati Prabhakar bristles. An accomplished scientist herself, normally reserved and measured, she becomes passionate when describing the intensity with which her scientists pursue technological surprise in aiming for military dominance. DARPA scientists are recruited from the best in their fields and, unlike other government employees, are limited to relatively short, four-year assignments, during which they recruit and fund other top scientists to push the boundaries of their disciplines. They work in numerous areas, from large strategic systems to small tactical ones, focusing on cyber technology, biology, neuroscience, and more. Some failure is expected, but the successes are numerous and generally extraordinary.

The ability of the United States to successfully implement its Third Offset strategy will depend on its innovative use of information technologies, autonomous systems, robotics, cyber techniques, nonlethal weapons, and high-speed weapons.

Information technologies are critical to artificial intelligence techniques, the autonomous military systems I mentioned earlier, robotics, advanced prostheses, and cyber weapons. They are also the basis for intelligence applications, such as data mining. The term "information technology" is a broad one, encompassing a range of areas including microelectronic devices such as microprocessors and transistors, the Internet, high-performance computing, algorithms, data collection and storage, data transmission, data mining and analysis, and the "Internet of things"—a term used to describe the growing trend of putting sensors on everything and tying them to the Internet. Computing power and memory have grown astronomically, as has the amount of data generated. The number of transistors that can now be placed on a chip lies in the hundreds of billions. The amount of data now gathered in two days exceeds all of the data created from the dawn of civilization to 2003.

The Internet of things has grown enormously, with the number of connected devices worldwide predicted to be almost forty billion in 2020. Appliances, vehicles, and even toys are connected. Unsurprisingly, the military has its own Internet of things. For several years, the Army issued helmets with built-in sensors to help diagnose brain injuries. Even individual munitions have Internet addresses. Almost everything—conventional bombs and nuclear weapons, soldier communications and satellites, simple vehicles and advanced armor systems, lasers and navigation systems—depends on computers and their components.

The Internet of things and Wi-Fi-enabled devices, with all of their advantages, also come with some serious downsides. Always-on sensors are threats to privacy and civil liberties, and the Internet of things was recently determined to have provided the hardware basis for a massive denial-of-service attack on East Coast Internet service providers. Information technology is central and crucial to military planners. Continued advances in it could render communications security obsolete. So-called quantum computing, if fully realized, will make cryptography impossible, allowing a quantum computer to break any code. Imagine a world, especially a military one, in which nothing can be kept secret. The enormous size and complexity of software systems will make understanding them difficult and may lead to unsafe assumptions about their provenance.

Such luminaries as Bill Gates, Elon Musk, and Stephen Hawking have sounded alarm bells over the continuing push for artificial intelligence. While the media often raises fears of apocalyptic scenarios with intelligent robots taking over the world, their concerns also include the more prosaic. They, too, worry about artificial intelligence systems going beyond humans' ability to control, but also include in their concerns such issues as how to determine system trustworthiness, and how to detect errors in algorithms. Additionally, they voice broader concerns about the societal impacts of ubiquitous artificial intelligence systems.

Massive data collection has been a boon to surveillance for law enforcement and intelligence agencies. A former chief technology officer for the CIA, Gus Hunt, said in an open forum a few years ago that the government was at a point where it could *conceivably* collect and analyze all of the information generated on everyone—within the limitations of the law, of course. This is important in the search for bad guys. However, ethicists and privacy advocates express a growing concern that data mining and big data are significant threats to personal privacy and offer at best only a snapshot of a situation. There is also legitimate unease over the data mining software and algorithms themselves. In my discussions with several intelligence agency analysts, they admitted that they had little or no detailed knowledge of the inner workings of some of the software they used, which was often provided to them by contractors. The question is whether or not there may be biases unknowingly built into the systems.

History has demonstrated that in times of crisis, the demand for greater surveillance and intelligence inevitably increases, along with an accompanying contraction of individual privacy and civil liberties. During and after the nation's wars, fear of foreign agents and disloyal Americans led to greater information collection and restrictions of many people's personal freedoms. Technological improvements in this area are insidious, as advances in sensing, communications, and information processing just invite greater surveillance. As sensitivity of sensors improves, the urge to collect ever-weaker signals, thinking we will uncover new information, is too great to ignore. As storage capacity increases and costs decline, the desire to collect more and more data points is irresistible. We collected data about every American telephone call because we could. Also, as we have seen with the unending reports of data theft, from Target Corporation to the Office of Personnel Management, security of information is always a primary concern.

Synthetic biology is an emerging area of research that can broadly be described as the design and construction of novel artificial biological pathways, organisms, or devices, or the redesign of existing natural biological systems. It holds great promise for new drugs, materials, and fuels. It may also lead to the development of new and dangerous organisms with as yet unimagined characteristics. Synthetic biology clearly has the potential to provide new physiological functions and improved performance.

I talked recently with Dr. Justin Sanchez, director of

DARPA's Biological Technologies Office, which invests in both biology and neuroscience programs. In creating the office several years ago, DARPA announced that it would focus on "discoveries that help maintain peak warfighter abilities." The agency indicated that its work would extend well beyond medical applications to include the exploration of complex biological issues that can affect a warfighter's ability to operate and interact in the biological and physical world. In the biology area, Dr. Sanchez's office supports work in rapid detection of pathogens, prediction of their mutations, and creation of countermeasures against them, as well as programs for quickly scaling up vaccine production.

The prime motivation for DARPA's biology work is the protection of the individual soldier. However, these technologies also provide the opportunity to undertake sophisticated intelligence, surveillance, reconnaissance, and offensive actions on adversary forces, such as intelligence gathering based on genetic profiles, or conducting tagging, tracking, and locating missions using biomarkers. While serving in the government, I was aware of concept feasibility studies on how a so-called naked human (one who is not creating electronic or other types of signatures) might be detected and tracked, or even tagged. Some of the ideas were biology-based, and suffice it to say that some of them—none of which went beyond a feasibility analysis—were pretty outrageous, and sometimes downright scary. In my conversations with Sanchez, he was adamant, and convincing, that their work in biotechnology was all for therapeutic or defensive purposes, but he reluctantly agreed that such advances in knowledge can just as easily be used for evil.

So-called gene editing techniques, like virus manipulation and the newly discovered CRISPR/cas9, are becoming more prevalent. The techniques are exciting in that they promise to help in eradicating serious diseases. On the other hand, such techniques may be used against us. So worrisome is the CRISPR capability that the director of national intelligence has listed it specifically as a bioterror threat. In an interview with Stanford University writer Mark Shwartz, biophysicist Steven Block raised the specter of "black biology," a shadowy science in which microorganisms are genetically engineered to create novel weapons. "If anthrax, smallpox, and other 'conventional' biological agents aren't frightening enough," Block says, "genetic maps of deadly viruses, bacteria, and other microorganisms already are widely available in the public domain." He goes on to talk about "stealth viruses" that could infect the host but remain silent until activated by some external trigger. Some scientists believe it is possible, using technologies similar to those used in developing individually customized genetic therapies, to assassinate high-value targets through similarly custom-designed viruses.

Synthetic biology has numerous potential dangers, among them environmental and safety risks, the possibility of the release of dangerous microbes into the environment, and the displacement of important natural microbes. According to the Department of Defense, "high performance and predictable operation of engineered biological systems is still a challenge." Also of concern are new adversary threats like the production of neurotoxins or infectious viruses resistant to vaccines and antiviral medications. Indeed, in a widely reported case, Australian scientists attempted to modify the mousepox virus to stimulate the production of viral antibodies in mice. The effort was intended to eradicate a disease that was ravaging the mouse population. The experiment failed to produce the desired antibodies. It did, however, result in a mousepox strain of extraordinary and unexpected lethality. Researchers found that even the mice vaccinated against the disease died. Fortunately, the experiment was conducted in a secure laboratory, where the problem was contained. This type of research on humans must be approached with extreme caution.

Many researchers are concerned about so-called gain-of-function experiments wherein genes are modified to do new things. While the United States and other Western countries prohibit such work, Chinese researchers have recently become among the first to apply the CRISPR gene editing technique to human embryos. At the same time, the Chinese company BGI (formerly Beijing Genomics Institute) is conducting large-scale gene sequencing studies of very-high-IQ individuals. Reportedly, Chinese researchers seek to use genetic engineering to increase the average intelligence in the population. Researchers in other countries have even conducted experiments in attempts to artificially create life. Such work raises prickly ethical, philosophical, and religious questions.

While there are new and more efficient techniques, gene manipulation and virus modification have been around for years. Colonel (Dr.) Matt Hepburn, one of the DARPA biology program managers, talked with me at length about his work on trying to predict and react quickly to virus mutations. If the work is successful, U.S. military forces operating in far-flung corners of the globe will have the ability to quickly diagnose virus-borne illnesses, develop treatments, and rush them into the field. Like his office director, Dr. Hepburn was unequivocal about the fact that DARPA's work, by policy and by law, does not include virus modification research. He was refreshingly willing, however, to speculate with me about the malicious ways in which work such as his could be used. Some of these I have described above. He spoke candidly about his career-long debate with himself over the dangers versus the health benefits of his research and obviously has concluded that the benefits outweigh the risks. It is good to know that he and others like him take the issue very seriously. DARPA's deputy director, Dr. Steve Walker, told me that every time he is briefed on new advances in biology research he comes away somewhat shaken by the implications of it all.

The U.S. intelligence community also has its version of DARPA, known as the Intelligence Advanced Research Projects Activity (IARPA). The agency's director, Dr. Jason Matheny, is a mathematician with field experience in epidemiology in underdeveloped countries. He is also a student of moral philosophy with an abiding interest in technology ethics. IARPA, too, supports research in synthetic biology directed, naturally, to the prediction of future threats to the United States. Dr. Matheny also expressed to me his serious ethical concerns about the conduct of leading-edge biology research, describing the intense oversight his agency provides.

Of further concern are the increased numbers of people—of all motivations—who are able to engage in synthetic biology. In fact, the International Genetically Engineered Machine (iGEM) Foundation conducts programs and international competitions for high school students. There is good reason for concern. Prediction of biological behavior is very hard, biological complexity is not well understood, and the fundamental "laws" of synthetic biology are not yet fully comprehended. Theoretical physicist Freeman Dyson, musing about the dangers of such broad access to synthetic biology, has said, "Biotech games, played by children down to kindergarten age but played with real eggs and seeds, could produce entirely new species—as a lark." Dyson further noted, "These games will be messy and possibly dangerous. Rules and regulations will be needed to make sure that our kids do not endanger themselves and others."

Neuroscience technologies are helping wounded soldiers recover from traumatic brain and other injuries. Rapid advances have been made in neural control of prosthetics and in the broader general area of brain/machine interfaces. Research is ongoing into the use of chip implants to restore brain functions and to stimulate the peripheral nervous system. Research at DARPA has shown that it is possible to implant a chip into a brain and have it download the signals from some specific brain activity. If brain function is subsequently destroyed, it can be restored by uploading the contents of the chip back into the damaged brain. Experiments conducted on nonhuman primates are now being extended to human trials. The goal of DARPA's new program is to develop new closed-loop systems that leverage the role of neural "replay" in the formation and recall of memory, to help individuals better remember specific episodic events and learned skills.

Other researchers have now shown that animals can collaborate via a brain-to-brain interface. As reported by biotechnology editor Susan Rojhan in the *MIT Technology Review,* pairs of rats have demonstrated that they can communicate through brain chips and collaborate to perform a task. Scientists at Duke University trained pairs of rats to press the correct lever when an indicator light above the lever switched on, to obtain a sip of water. They next connected the two animals' brains via microelectrodes. One animal received a visual cue for which lever to press in exchange for a food pellet. When it pressed the correct lever, a sample of its brain activity was translated into an electrical stimulation that was delivered directly into the brain of the second animal. The second rat had the same types of levers, but it did not receive any visual cue for which lever it should press. To press the correct lever and receive the reward, it had to rely on the cue transmitted from the first rat. The second rat ultimately achieved a statistically very high success rate. When it did commit an error, the first rat basically changed both its brain function and behavior to make it easier for its partner to get it right. In a second set of experiments the researchers trained pairs of rats to distinguish between a narrow or a wide opening using their whiskers. During trials, the first rat detected the opening width and transmitted the choice to the second rat, which achieved a success rate significantly above chance. To test the transmission limits of the brain-to-brain communication, the researchers placed one rat in Brazil and transmitted its brain signals over the Internet to a second rat in North Carolina. They found that the two rats could still work together on the tactile discrimination task.

The military implications of these advancements are profound. The technology is absolutely in its infancy, and the amount and type of data that can be transmitted is currently limited, but this will inevitably improve. Covert communication, always a goal of intelligence operatives and special forces, is but one possibility. Imagine a scenario in which a soldier on a mountaintop could communicate with a fellow soldier in a valley to warn him of an impending ambush—without having to give away his position. Even if an adversary intercepted the communication, what would they hear? Better yet, if not two but multiple brains can be linked together and learn together, the result would be tantamount to massively parallel computing. It would certainly give new meaning to the term "brainstorming."

DARPA has demonstrated the utility of neuroscience for intelligence operations in a project called Neuroscience for Intelligence Analysis (NIA). In this project, a soldier's brain was monitored and recorded seeing a target in a scene significantly earlier than he could press a button to say he'd seen it. In the laboratory, an analyst's brain activity is sensed using electroencephalographic (EEG) techniques that have been able to detect a fast, earlier signal in the brain associated with the detection of targets in imagery. This finding suggests that it may be possible to extract target detection signals from complex imagery in real time using noninvasive neurophysiological assessment tools. This implies that the brain could be wired directly, bypassing other, slower bodily functions. For now, researchers still feel that the human visual system is the most accurate target detection apparatus, but its augmentation with neuroscience-based measurement capabilities could dramatically increase the speed and accuracy of the human analyst.

While invasive brain research work to date has been aimed at therapeutics, and performance research has used data gathered from the brain itself, scientists are taking the next logical step. DARPA recently announced a project to enhance the abilities of healthy soldiers to learn, using external stimulation not of the brain, but of the peripheral nervous system, which consists of the nerves and ganglia outside of the brain and spinal cord. Quoting DARPA, the new program, Targeted Neuroplasticity Training (TNT), "seeks to advance the pace and effectiveness of cognitive skills training through the precise activation of *peripheral nerves* that can in turn promote and strengthen neuronal connections in the brain." If successful, TNT could accelerate learning and reduce the time needed to train foreign language specialists, intelligence analysts, cryptographers, and others. The program is notable because, unlike many of DARPA's previous neuroscience and neurotechnology endeavors, it will aim not just to restore lost function but *to advance capabilities beyond normal levels.* The last point is hugely important. It is not surprising that the Defense Department has taken this step, but what is surprising is that the reaction from ethicists and moral philosophers has been so muted. As is true for so much that lies on the horizon, this cries out for public discussion.

Dr. Doug Weber is the program manager for the TNT effort. He has been involved in much of the advanced prosthetics work, developing artificial limbs that give the wearer sensations of touch. The control devices, implanted in the damaged limb and the prosthetic, are octopuslike, with a central processor and numerous tiny electrical leads, like tentacles, that attach to nerves. Weber explained to me that not only can the targeted neuroplasticity technique be used to assist in learning, but stimulating the peripheral nerves has achieved success in treating post-traumatic stress disorder.

That researchers are able to access the brain through peripheral nerves is enormously significant.

Dr. Weber agreed that eventually signals could be designed for specific functions and could also be transmitted wirelessly over long distances, in effect getting inside a soldier's brain from afar. Since these techniques involve data and communications, they could then become a target of cyber warfare. While admittedly speculative, the military and intelligence implications of such developments are noteworthy, and probably limited only by the imagination.

Neuroscience raises numerous ethical issues, despite miraculous breakthroughs such as the development of mind-controlled prosthetics. Researchers are now able, in a sense, to "read the minds" of individuals. Of course, as Dr. Sanchez pointed out, defining what is meant by "the mind" is fraught with ambiguity. Be that as it may, if we can read thoughts, whatever that means, it also stands to reason that we can write them too. Researchers know enough about the function of neurons and their signaling to create mind-controlled prosthetic limbs with artificially induced "feeling," in which an external signal is sent back into the brain. Once the structures of thoughts and emotions are understood, it is perfectly conceivable that they could be inserted into the brain from an external source.

Neurological enhancements raise deep ethical questions and concerns about compromised character, justice, and coercion. What effects do neurological enhancements have on "human essence"? What will be the effect on personal autonomy and free will? How do we determine which soldiers get the enhancements? They will, of course, pose enormous challenges for privacy. The security and proper use of the data becomes a real concern. What would we do if in fact we found significant brain differences among populations? Could the information be used as a rationale to discriminate? Some writers are already beginning to sound the alarm about a return to the eugenics movement of the early twentieth century. Among other worrisome issues for neuroscience are its possibilities for use as a tool for unreasonable search or its potential for misuse in interrogations. Finally, if neural implants are used or if other neural enhancements are made to a soldier, what happens when that soldier transitions back to civilian life?

Information technology, synthetic biology, and neuroscience all contribute to the broad area of enhancements. The same technologies that might restore an individual's ability to a species norm could also be used to bring someone above that norm. For the military, this implies enhanced soldiers. External enhancements might include exoskeletons, ballistic or radiation protection, and advanced vision devices like contact lenses that would allow a soldier to see in the infrared spectrum, enabling night operations without the burden of bulky helmet-mounted gear. Internal enhancements might include pharmaceuticals, genetic modification, biological or metabolic changes, or the surgical implantation of computer chips into the brain.

Super soldiers have long been the stuff of science fiction. Television shows like *The Six Million Dollar Man* and movies like *Iron Man* and *RoboCop* allow us to imagine what would happen if humans could be augmented with superhuman capabilities. While much of what these depict remains the stuff of science fiction or even scientific impossibility (antigravity suits, for example), a substantial percentage has become reality or is being researched. The Defense Department has been working on these technologies for a decade. Past projects such as the "Metabolically Dominant Soldier" and "Soldier Peak Performance" studied biological, genetic, and metabolic ways to improve battlefield performance. The technologies would allow for rapid tissue regeneration, faster healing, greater muscle strength, cognitive enhancement, the ability to operate without sleep for many days without performance degradation, higher metabolic energy, and immunity to pain.

The deputy secretary of defense, Robert Work, recently said about Russia, "In at least one area, our adversaries are ahead: enhancing human performance by modifying the body and brain itself. Now our adversaries quite frankly are pursuing enhanced human operations and it scares the crap out of us, really. We're going to have to have a big, big decision on whether we're comfortable going that way."

Warriors are enhanced, and growing ever more so. The soldier of yesterday carried around his helmet, his weapon, ammunition, some food, and maybe a tent to sleep in. If he was lucky, his leader had a radio. The soldier of today has all those things and more. He must also wear his protective armor vest and carry night-vision goggles, his handheld computer, radio, and batteries to power everything. The soldier of the future will have all of these things, many of them made smaller and lighter for him, but he will also be instrumented, monitored, and analyzed much as NASA has always done with its astronauts. The soldier of the future will be a collection of data points.

The U.S. Army program known as Future Soldier addressed how the 2030 soldier would be equipped. It included designs for human performance, protection, lethality, networking, and sensors. The program placed special emphasis on cognitive performance to improve effectiveness and operational tempo through technologies such as cognition-enhancing drugs, physical enhancements, and neural prosthetics, or brain chips. Monitoring included behavioral, neural, and internal information, such as blood oxygenation and glucose, all linked to an individual predictive computer model of soldier performance. Much of this work continues today in different ongoing Army research projects.

The above technologies are all intended to keep the soldier safe on the battlefield and to provide him or her with that extra advantage that might mean the difference between victory and defeat. However, some researchers are concerned about providing soldiers drugs to forget or to decrease fear or inhibitions. The bioethicist Jonathan Moreno has pointed out that philosophers believe the idea of ourselves is intimately bound up with our memories, and that "anyone who believes that there are certain boundaries that should not be crossed must be concerned about the modification of the ability to remember and to forget." These emotions have value in keeping the soldier safe and in moderating his behavior. Author and former Marine Karl Marlantes has said that "remembering our common humanity and controlling the beast that wants to obliterate that memory is the task for all conscious warriors of the future."

What of freedom and autonomy? Is a soldier operating in an enhanced state really operating of his or her own free will? Can a pharmaceutically enhanced soldier be considered a conscious one? Can he be considered morally responsible for his actions? Should the soldier even be allowed to leave military service once enhanced? What will be the effect on the local civilian community of an enhanced former soldier? What about the implications of an enhanced-soldier arms race? Will enhancement technology proliferate, and if it does, what will be the effects on the civilian community if criminals are able to access it? There is a great deal of uncertainty and risk in the development and employment of these technologies, and this risk must be considered and debated before they are placed into operations. We must understand the long-term and perhaps unintended effects of these technologies on the soldiers themselves and their units. We must answer open questions about the creation of dependencies and addictions, or whether certain enhancements are permanent or reversible. While the enhancement of soldiers cannot a priori be ruled out as illegal or unethical, there must be debate about questions of military necessity, legitimate purpose, and warfighter dignity, safety, and accountability.

Robots have long captured the popular imagination. They can be "good" robots like the lovable WALL-E or the engaging and comical robot in *Short Circuit,* or they can be "bad" robots like HAL in *2001: A Space Odyssey* or the fearsome, murderous robots in *War of the Worlds.* These are fictional, but real robots become more capable every day with improvements in electronics, computers, artificial intelligence, materials, and other supporting technologies.

Robots can be human in form or they can take on the most efficient form for the purpose and the situation in which they will operate. They can be disembodied computer systems, like the alluring operating system in the movie *Her,* or they can be autonomous unmanned aerial or undersea weapons. They can be large, like Russia's purported nuclear-weapon-carrying autonomous submarine, or they can be as small as an insect. Robots can be controlled, semiautonomous, or autonomous. They can also be lethal.

Drone strikes, both military and otherwise, are commonplace. Soon the drones will be able to make engagement decisions on their own. Unmanned aerial, terrestrial, and naval systems with new, potentially autonomous and lethal capabilities are now being developed, and can be controlled by artificial intelligence. Many such systems already exist.

On the eastern bank of the Potomac River, just outside of Washington, D.C., sits the U.S. Naval Research Laboratory. Formed in 1923, the laboratory has been responsible for many of the advanced technologies used in the military today, especially in such areas as communications and undersea warfare. Today, work is ongoing in space systems research, tactical electronic warfare, microelectronic devices, and artificial intelligence. A new addition to the laboratory is a massive new facility dedicated to research in unmanned vehicles, known as the Laboratory for Autonomous Systems Research. The building is cavernous and outfitted with hundreds of cameras and sensors to monitor the behavior of unmanned vehicles being tested. During my visit, I watched multiple "quadcopters" operate in unison, ostensibly cooperating in a search for some hidden target. I entered a large hot, humid room, densely overgrown with tropical plants, where I was being surveilled by covert autonomous systems apparently designed specifically for such environments. And of course, since I was in a Navy facility, I watched as unmanned swimming vehicles were tested in the giant pool. In a similar facility at the Naval Postgraduate School, I watched as Navy divers tested underwater drones intended to act as team members with other divers, autonomously taking on tasks too dangerous for humans. The Navy is even researching the possibility of "flimmers," vehicles that can operate autonomously both in the air and under the water.

At the sprawling Eglin Air Force Base in Florida, researchers developed the Low Cost Autonomous Attack System (LOCAAS). Envisioned as a miniature, autonomous munition capable of broad area search, identification, and destruction of a range of mobile ground targets, it was designed to loiter over a battlefield, spot a target, decide on its own, and attack. While the system worked, it was never placed in operational service. The military was not quite ready to take that step.

As late as 2011, the U.S. Air Force Research Laboratory was developing both semiautonomous and autonomous bird-size and insect-size lethal drones. In concept, they would be employed just like a mini-LOCAAS, loitering aloft around an area and, sensing what they believed to be a legitimate target, delivering a lethal blow, explosive or otherwise, on their own cognition. At Eglin AFB there is also a laboratory devoted to the area of biomimetics, or the science of trying to make man-made systems mimic the behavior of natural ones. When I visited the lab, researchers were painstakingly measuring the electrical output of each element of a fly's eye with microscopic electrodes and modeling the behavior of the eye as objects passed in front of it. The military interest, the researchers told me, is to know how the fly so successfully evades threats. Other experiments included studying dragonflies to learn about their flight characteristics, and high-speed photography of bees, which can continue extremely stable flight in turbulent conditions.

Since 2012, stated DOD policy is that the military will not employ completely autonomous lethal systems. Unfortunately, the department has issued no further guidance or implementing instructions since the publication of its directive, and the moratorium on lethal autonomous weapons is only to last for ten years. Notwithstanding the stated policy, military strategist and author Thomas Adams described the situation thirty-five years from now as one in which "small, lethal, sensing, emitting, flying, crawling, exploding, thinking objects may make the battlefield highly lethal."

Robots and autonomous systems allow us to reduce the risk to our warfighters. Unmanned systems are valuable to military planners because they allow operation over great distances, in inaccessible areas. They do jobs with more efficiency and at lower cost than soldiers and, most important, can engage in lethal operations without exposing soldiers to injury or death. Robots and unmanned systems are force multipliers, giving commanders the capability to employ greater power with fewer personnel. They expand the battle space, by going places where humans cannot go, or cannot safely go, and at distances unattainable by humans. They do not get emotional, and thus reduce or eliminate unethical conduct by soldiers.

Debate rages, however, about whether autonomous robots can do a better job of minimizing bad behavior and unnecessary casualties, and whether they should. In 2006, the U.S. Army surgeon general published a surprising and troubling report about the behavior of soldiers in the Afghanistan and Iraq theaters of war. The report found that many soldiers either ignored the rules of war, or said that they would do so, under the stress of combat. Georgia Tech University professor Ron Arkin claims that robots can be programmed to act better. Other experts disagree, claiming that it is impossible to program into a robot all of the nuanced thinking and decision making required of soldiers on a battlefield. Skeptics worry that machines will be incapable of empathy and question whether they will be more or less humane than humans. Still others, like myself, remain opposed to the mere idea of a machine having lethal decision-making authority over a human life, regardless of its capabilities.

There are numerous and strong objections to giving autonomous systems lethal capability. There is even a global effort to create a ban on such systems. Underlying many of the concerns is a skepticism that any computer program could satisfy ethical and legal principles, a Kantian belief that it is categorically wrong to take the human moral agent out of the firing decision, and that by removing risk the incentive to use armed force rises.

Other issues surrounding the research and potential employment of autonomous systems, especially lethal autonomous weapons, are complexity and unpredictability. These systems contain millions of lines of computer code, and computer programs inevitably contain errors. Some of these may be fatal. Also, taking humans out of the decision-making process undermines the possibility of holding anyone accountable, a key element of military operations and ethos. And what happens if nonrational actors get their hands on these systems?

All of the autonomous lethal robotic weapon systems currently in use are defensive in nature. Examples include the Navy's Phalanx ship defense system and the Patriot air defense system. Robotic systems used for offensive purposes remain, for now, operator-controlled. The Department of Defense says that for the application of lethal force in combat there will always be a man in the loop as an integral component in the decision to take action, or in a position where he can intercede in a machine's decision process. However, there are great pressures to move toward autonomous weapons. The demand for budget reductions and the drive to downsize the military will make robots a more attractive alternative. Pentagon planners are quoted as saying that they need systems untethered from human masters, that robotic systems can save money, and that the main current issue is cost/benefit analysis. Machines are, or will be, relatively cheaper than soldiers. Will the lower cost of entry for this technology drive more widespread use?

In an interesting psychological twist, sometimes it doesn't even seem to matter whether there is a man in the loop. In 1988, the U.S. guided missile cruiser *Vincennes,* operating in the Strait of Hormuz, shot down an Iranian airliner, killing everyone on board. Tensions were high that day and the system was operating as designed. When questioned in a 2000 BBC documentary, a U.S. government spokesman said the incident may have been caused by a psychological condition among the crew of the *Vincennes* called "scenario fulfillment," said to occur when warfighters are operating under combat stress. In such a situation, the crew will carry out a training scenario as if it were real, while ignoring information that contradicts the scenario. In this incident, the scenario was an attack by a lone military aircraft. In tragic fact, the aircraft was not military at all.

As robotic systems take on greater roles, the risk of interruption of command links becomes a greater threat and systems will have to be able to operate on their own without communicating with headquarters. This is especially true now in an era in which cyber warfare and electromagnetic warfare are on the rise. As the complexity and speed of systems and operations increase, the rate of human decision making will become a limiting factor. Humans will simply be unable to keep up with the speed of battle.

Air Force chief scientist Greg Zacharias recently told a military group that "as early as the 2020s, the Air Force could implement the 'wingman' concept—fully automated drones that would fly alongside manned aircraft." He went on to say, "The Air Force Special Ops guys are working with an unmanned vehicle that will launch from their C-130 [cargo aircraft] and can go down under the clouds to take a look at what's going on, without risking the C-130 and its complement." In an even more technically aggressive project, DARPA is working on a concept called "Gremlins." The program envisions launching groups of unmanned aerial vehicles from large aircraft. When the vehicles complete their mission, a transport aircraft would retrieve them in the air and carry them back to base, where ground crews would prepare them to be used again. This is not as far-fetched as it sounds. In the 1960s, pilots routinely flew specially equipped transport aircraft to retrieve, in-flight, film canisters that had been dropped to earth from satellites.

In their article "20YY: The Future of Warfare," Paul Scharre and Shawn Brimley tell us that swarms of networked, autonomous systems could operate with greater coordination and speed of maneuver than is possible with human-controlled systems. They describe the swarms as saturating and overwhelming enemy defenses and, being low-cost, soaking up enemy missiles at favorable cost-exchange ratios. These swarms, they say, "could jam, spoof, and disable enemy sensors, sowing confusion and raising the electromagnetic noise level to hide follow-on U.S. strike platforms."

Already we have seen Russian-backed forces in Ukraine take advantage of swarm behavior. Rebels, in addition to effectively employing tanks and artillery, have incorporated unmanned aerial vehicles in swarms to conduct cyberattacks and interfere with battlefield communications and GPS guidance systems.

Military planners wishing to incorporate autonomous robots will need to be concerned about the age-old concept of unit cohesion, an important indicator of performance. Will a human/robot team be able to function smoothly? Consider a case in which a human/robot team is conducting an operation with a human in command. Will a robot be able to refuse an unlawful order? Or if a robot sees and reports on everything, will it affect the behavior of the other soldiers? And how will one side react when one of its robots is captured and the same technology is used against it? By their very nature, autonomous systems do not require operators with experience making life-and-death decisions. Army major Daniel Sukman pointed out that "removing a level of operators who live in the world of tactics may remove a cohort which needs that experience when they become operational and strategic leaders."

Newspapers are filled with reports of cyber intrusion, cyber theft, hacking, or mischievous activity. Some of these events cause havoc in the civilian realm. Military planners discuss cyber operations that include defense, exploitation, and attack, and the words have important meanings. All advanced militaries are becoming increasingly dependent on computer and communications technology. Cyber weapons thus provide an important means of destroying, degrading, disrupting, and denying adversary systems.

Years ago, I was responsible for building some of the Air Force's most critical command and control and war planning systems. At the time we were just beginning to understand the vulnerability of our systems to cyberattack (and, conversely, the opportunities such attacks presented for ourselves). On a visit to a secure facility deep in a New Mexico mountainside, long before cyber warfare became a hot topic, research scientists demonstrated to me the ability to remotely and anonymously disable or destroy a target computer, and to turn industrial switching control systems on and off. Such techniques are well known and commonplace today, but the ability to do them covertly remains a closely guarded secret.

The detection and thwarting of cyberattacks is of enormous importance. While the struggle to stay ahead of determined adversaries continues, the National Security Agency and the U.S. Cyber Command have to date successfully defended critical national security networks from intrusions and attacks.

In cyber operations, the vulnerabilities of weapons platforms, as well as a nation's infrastructure, are exploited or destroyed. Such techniques loom large as expertise grows and proliferates. The ability to destroy equipment and possibly military capability was evident in the Stuxnet attack on the Iranian nuclear centrifuges. The ability to destroy information, reputation, and financial resources was evident in the North Korean attack on Sony Pictures. There have been instances of passengers hacking into aircraft flight systems from their coach seats, and it was recently determined that control systems for a dam in upstate New York had been penetrated by Iranian hackers.

Because of the speed at which cyber war will be conducted, the difficulty of distinguishing between exploitation and attack is great, and dangerous misperceptions can result. Is a massive denial-of-service attack from Russian hardware devices a prelude to a real attack, or is it merely the work of independent hackers through a Russian service provider? Operations at computer speeds in such a complex sphere of battle tax the capabilities of today's highly trained cyber soldiers, not to mention those of our decision makers.

Cyber warfare can potentially reduce the harm and physical damage of war. It provides yet another, largely nonlethal alternative to classic kinetic warfare and a means of preventing escalation into such conflict. However, it could replace physical damage with economic, digital, or informational harm. Concerns about cyber conflict include the degree of certainty in attribution needed for a cyber response, and the possible loss of control of cyber weapons that may escape and cause harm on a wider scale than intended. Worms, viruses, and other types of malware, once released, propagate much like they do in biological systems. Even if they are targeted at specific systems, they can infect others, as networks are vast and constantly reconfiguring themselves.

Uncertainty is greater in cyber operations because the technology allowing opponents to disguise their identity is so sophisticated. Is it ethical to attack when the identity of parties is uncertain? In classical combat and decision making, the identity of your opposing force is relatively well understood. Experts say that once a weapon is "released into the wild" it is difficult to predict exactly where it will propagate. The Stuxnet virus targeting the Iran nuclear program is a case in point. That software was looking across networks for a specific type of industrial controller and found its way onto systems all over the world. Even then, it was designed to damage only a specific configuration. While it did not damage any other systems, there was a cost, in time or money, to innocent users to have it removed. Is it ethical to use poorly controllable technology? How do we assure the proper type of attacks when damage assessment is so difficult?

The Department of Defense recently announced that it will issue contracts to industry for the development of lethal cyber capabilities. In a direct analogy with military planning for conventional conflicts in which plans are made to fire explosive munitions, the department has instructed the appropriate units to create plans to fire "cyberspace munitions." Lethal cyber munitions might, for instance, be designed to interfere with the flight controls of an aircraft or to cause a bomb to detonate prematurely. Lethal cyber weapons are important and frightening additions to the tools of war, the ramifications of which are huge and largely unknown. We must think deeply about the consequences of using such weapons. There are also many as yet unanswered questions about the uses and methods of deterrence, and means of conflict deescalation once a cyber war begins.

The Department of Defense has decreed that cyber warfare, like other forms of warfare, must adhere to laws of armed conflict. It must be aimed at harming military infrastructure, must degrade the adversary's ability for kinetic war, do minimal harm to civilian life or property, and be a last resort, after negotiations have failed. However, the chief of the U.S. Army's Cyber Center of Excellence, responsible for teaching Army cyber warriors the ethical guidelines under which they are to operate, worries that "we are asking a lot from our young soldiers who are entering a world of lightning fast decision making and a very complex and cloudy environment."

Conflict in the electromagnetic spectrum, specifically in the radio-frequency domain, is a related but distinctly different issue. All military command and control systems and almost all weapons now use advanced electronics in some fashion. William Forstchen's 2009 novel *One Second After* depicts the nonblast, electromagnetic-pulse effects of a nuclear explosion on the population of the United States, an event that instantly disables almost every electrical device in the country and throughout the world. The world is plunged into darkness and chaos.

Fortunately, there are other, less destructive and more targeted ways to jam or destroy electronic systems. The U.S. Air Force designed and built a munition that when detonated above a target creates not an explosion with fire, intense pressure, and steel fragments to cause damage, but a high-power radio-frequency pulse that will burn out or damage the functions of all electronics within a given radius. Such techniques and systems are closely held secrets, as knowledge of their designs and capabilities would cause an adversary to develop effective countermeasures. For reasons of safety as well as security, they are tested in isolated locations, away from prying eyes and sensors. Because much of my work in the military was highly dependent on advanced electronic systems, I was keenly interested in them. In one memorable incident, I was helicoptered to a remote part of a desert base in the U.S. Southwest. I was told to pay close attention to all of the electronic systems, which moments later flickered and died. That was it. No bang. No flash. Demonstration over. Such is the world of future warfare.

In many situations, including peacekeeping, crowd control, antipiracy, and humanitarian missions, it would be desirable for U.S. forces to have operational options other than the use of deadly force. Nonlethal weapons like chemical, audio, electrical, laser, and radio-frequency devices offer viable options in appropriate scenarios. Some ships are equipped with systems that generate unbearably loud sonic waves to deter pirates. Military checkpoints are equipped with dazzling, nonblinding lasers. These systems are intended to inflict pain, immobilize, and otherwise render an enemy incapable of combat, but are not lethal.

In an era in which there is an increasing reluctance to accept war deaths, nonlethal weapons offer an attractive alternative, especially because in today's conflicts, combatants and noncombatants are deliberately mixed. Consider, for example, Hamas's use of civilians as human shields, or the increasingly frequent terrorist hostage-taking situations.

Nonlethal weapons, however, concern ethicists and military planners alike. Many intentionally do not discriminate, meaning that they affect all people in a particular area. In 2002, Chechen terrorists took numerous hostages in a Moscow theater. Russian special forces decided that rather than storm the theater and risk massive casualties, they would inject a calmative drug into the ventilation system. Unfortunately, the drug as administered was fatal to a large number of the hostages as well as the attackers.

Some systems can aim a pulse of energy at a vehicle and stop its engine. Biological agents can attack rubber or metal components in vehicles. One system, designed to be used against personnel, ostensibly as a crowd dispersal device, is the Active Denial System, commonly referred to as the "pain ray." The Air Force Research Laboratory developed the device, a truck-mounted system with a high-energy radio-frequency microwave source and a focusing antenna. The beam instantaneously affects only the very top layer of skin, heating the water therein and causing extreme discomfort but no lasting injury. While very effective, the system has never been operationally deployed, largely because the rules of engagement for such a device remain elusive. What if its targets are prevented from moving by, say, a large crowd? What about the presence of children?

After 9/11, force protection and the security of military bases became even more important to senior leaders. In an attempt to understand the Active Denial System's possible uses in base defense, I toured the laboratory during its early development and was given an opportunity to fire the weapon or to be a target. As I looked through the rangefinder and pulled the trigger, I expected to hear a zap or a whir—something. There was nothing but the hum of the system electronics, the air conditioner in the van, and the sight of the volunteer targets downrange running for cover. It is a powerful and effective device.

Laser weapons are moving quickly from the realm of science fiction and laboratory curiosity to actual military platforms. For years, the Air Force has studied and tested high-powered chemical lasers on board large aircraft for defense against ballistic missiles. More recently, the service successfully tested a smaller chemical laser on board a more moderately sized C-130 aircraft and demonstrated that it could disable ground targets the size of trucks. The Navy has recently installed a laser system on the USS *Ponce* as a defense against threats from both sea and air. The Army is also testing mobile ground-based lasers for base defense against mortar and drone attacks. The laser is silent and invisible to the naked eye. The only sound you hear is the incoming aircraft or missile exploding due to the intense heat of the beam.

In a hypothetical forced entry into enemy territory, U.S. aircraft will probably not enjoy air superiority and will be subject to withering antiaircraft missile attacks. Air Force researchers are developing smaller, solid-state lasers for fighter aircraft to use in self-defense against these missiles. These devices use electrical power to excite solid lasing materials rather than requiring large reservoirs of liquids or gases. Such defensive systems are of course designed to detect and engage automatically.

The United States and other advanced nations have all experimented with large ground-based lasers. Because they are in fixed locations, are available in limited numbers, and their possible targets have such varied altitudes and orbits, their potential effectiveness as weapons are severely limited. They are, however, of great use in advanced optics for satellite detection and identification. At the Starfire Optical Range at Kirtland Air Force Base, New Mexico, I watched as scientists used a technique called optical conjugation to create crystal-clear images, through a turbulent and particulate-filled atmosphere, of satellites circling the earth. The chief scientist explained that a laser senses the atmospheric turbulence and guides hundreds of small mechanical pistonlike devices to deform mirrors in the system thousands of times per second to eliminate the distortion. Once a closely held secret during the Cold War, the adaptive optics are now a staple of military and astronomical systems alike.

After the end of the Cold War, America wanted a nonnuclear weapon that could nonetheless keep the strategic weapons of other countries at bay and, critically, could not be defended against. The militarily revolutionary field of hypersonic technology for both munitions and vehicle platforms has rapidly emerged to meet that need. These systems allow the using party to reach targets before the targeted party has an opportunity to defend or move them. The U.S. military is currently developing a hypersonic aircraft, which will progress alongside a hypersonic weapons program.

While today's cruise missiles travel at subsonic speeds, hypersonic weapons will be able to travel at five or more times the speed of sound. New hypersonic vehicles could be used to transport sensors, equipment, or weaponry. In a future war, volleys of missiles might scream toward an adversary country at ten to twenty times the speed of sound. Each missile would be aimed at a high-value target—nuclear missile launchers, military radars, submarine bases, or command and control centers. Within minutes, most of the opponent's strategic capabilities would be destroyed without the attacking party having to resort to nuclear weapons. The leadership would be blind, deaf, mute, and unable to respond. In the United States, such weapons would likely be hypersonic boost-glide vehicles, which ride atop a rocket and are released to attack their prey. Because these vehicles can maneuver, and because of their enormous speeds, shooting them down with conventional missile defenses is next to impossible.

High-orbiting satellites, using infrared technology, constantly monitor the surface of the earth and can detect the launch of a missile from almost anywhere. Inside the U.S. missile warning command center, displays instantly light up when a missile launch is detected, spurring teams into action. Almost always there is some advanced warning of the launch, or the systems identify the missiles as nonthreatening. In the Cold War, when we worried about nuclear attacks from the Soviet Union, we knew we had between ten and thirty minutes for the command authorities to decide what to do before impact, whether to shoot back or ride out the attack. With the threat of nuclear attack much diminished, we now face the question of whether the launch we just detected was a nuclear missile or a hypersonic boost-glide vehicle. Launched as they are atop large missiles, these vehicles have the very real and dangerous potential to be mistaken for ICBM launches and nuclear reentry vehicles. Can we waste time watching to see what it does, or must we respond immediately? Hypersonic missiles and aircraft, like high-powered laser weapons and computer attacks, will increasingly strain the capability of human operators to keep up. There will be very short timelines, creating greater dependence on machines for decision making. Fortunately, the young men and women in the command center are not faced with that time-critical decision, but our leaders will be. As will the adversary's.

Space is an ever-growing arena of importance for the military. The United States is heavily dependent on space systems, and our adversaries know it. Intelligence tells us that both China and Russia have researched and tested antisatellite systems. In the 1980s, the United States also tested an antisatellite weapon against one of our old satellites, firing a missile from a fighter flying at high altitude. China demonstrated its abilities when in 2007 it shot down one of its own satellites, creating an enormous cloud of space debris. In 2008, one of our nation's spy satellites failed after launch, and, citing concerns about ground safety, the United States conducted Operation Burnt Frost, during which a U.S. Navy system shot the errant satellite down over open water. U.S. military and intelligence officials remain concerned about space-based antisatellite systems, citing frequent cases in which foreign systems have come too close to ours. In July 2014, the Air Force launched two satellites designed to keep tabs on objects in geosynchronous orbit, home to the military's critical communications and missile warning satellites. Space is no longer the sanctuary it once was.

INCREASED COMPLEXITY

By any measure, the rate of technology development and adoption is accelerating and has brought with it incredible advances for society and for the military. Yet rapid technology growth itself poses an ethical challenge. *Time* magazine columnist Stewart Brand wrote in 2000, prior to the latest burst of technological innovation, that "change that is too rapid can be deeply divisive; if only an elite can keep up, the rest of us will grow increasingly mystified about how the world works." As he said, "We can understand natural biology, subtle as it is, because it holds still. But how will we ever be able to understand quantum computing or nanotechnology if its subtlety keeps accelerating away from us?"

Because they cannot or do not understand, people will simply give up trying. This is unfortunate and dangerous. Citizens and decision makers need to at least attempt to understand the technologies they are so eager to adopt and not be so willing to accept without question what technology companies offer. Ubiquitous personal electronics have already changed behavior in fundamental ways. What will be the societal impact of self-driving automobiles, or artificial intelligence? Do our choices have environmental consequences?

If the situation of dizzying complexity weren't already bad enough, we have been moving steadily into an era that computer scientist Danny Hillis refers to as "entanglement." As he defines it, entanglement refers to the fact that our systems are becoming ever more complex as well as ever more connected to one another. While the behavior of individual components may be computable and respond in predictable ways, that of the collection is too complicated to calculate. Human understanding of the whole system is no longer even possible, though we continue to try.

Finally, not only have our systems gotten exceedingly complex, but they end up being adaptive as well. Their behavior may change depending on feedback from the environment in which they operate. Typical examples of complex adaptive systems include the global economic network, the stock market, the brain and the immune system, human social group behavior, and terrorist networks. The Internet and cyberspace, a complex mix of human/computer interactions so fundamental to future weapons, is also a complex adaptive system.

Our adversaries, both peer nations and emerging entities, will take advantage of the complexity of the future battlefield. They will know full well the difficulty in understanding the behavior of some advanced technologies and weapons and will surely attempt to exploit that ambiguity by hacking our systems, injecting false data or otherwise causing us to question the veracity of our own, and thereby sowing seeds of uncertainty in our decision making. They too will have access to the technologies we are rushing to adopt, and they well understand our dependence on high-tech systems.

We can expect the future battlefield to be full of surprises, or, as former secretary of defense Donald Rumsfeld called them, "unknown unknowns." Are we prepared for the deadly dilemmas that will inevitably result?

2

HOW WE GOT TO NOW

AS A SWARM OF OUR DRONES *carefully navigates its way into the heavily defended country in preparation for the fighters and bombers that will follow, it passes over towns and villages that had not been expected to be impediments to its ingress. Suddenly, onboard sensors indicate air defense radars in large numbers placed among the population in farmhouses, shopping districts, and hotels. As the drones are individually targeted and destroyed by ground fire, command is seamlessly transferred among them. As one is destroyed, another automatically takes charge. Return fire is withheld in an attempt to avoid civilian casualties. Sensing that their effectiveness is being greatly diminished, the drone leader's computer attempts to seek authority from their human handler to engage, but is thwarted by heavy jamming. Unable to reach a human, it assumes decision authority, and, in an instant, the swarm of drones destroys all ground threats, ignoring their emplacement among the inhabitants. Total time of the engagement: ninety seconds. Meanwhile, the adversary's covert operatives release deadly, delayed-action pathogens into the air circulation systems of commercial aircraft from a dozen countries inbound to Washington, New York, Hous-*

ton, Chicago, Los Angeles, and San Francisco. The resulting epidemic creates chaos and panic among the people.

In 49 BC, Julius Caesar led his legions south over the Rubicon River en route to Rome. In doing so, he knowingly committed a capital offense against the will of the Roman Senate and made armed conflict inevitable. It is said that upon crossing the river, Caesar uttered the famous phrase *alea iacta est* ("the die has been cast"). The phrase "crossing the Rubicon" is now generally used to mean committing oneself irrevocably to a dangerous course of action. As technology surges ahead, and war is mediated more and more by computers and their machine embodiments, or enters the realm of biology and genetics, we may be approaching a Rubicon without even knowing it.

Planners and decision makers in previous ages were also challenged to comprehend the new conflicts and technologies of their time. Any reading of the history of war reveals that its evolution has always depended heavily on advances in technology of some sort, whether it be weapons, communications, or transportation. The introduction of gunpowder, the adoption of wireless radio, and the use of armored vehicles totally changed combat tactics. Nuclear weapons, of course, changed everything.

PAST WARS

The types of conflicts we have fought have evolved from the set-piece battles of the middle ages and the seventeenth and eighteenth centuries and the army-versus-army engagements of World War I and World War II to guerrilla wars, insurgencies, humanitarian interventions, civil wars, and the global war on terror, with its undefined battlefield and unpredictable timetable. Since the middle of the last century, global wars have been supplanted by an ever-changing and ever more complex set of conflicts involving not only nations, but also nonstate actors and interest groups. Some have termed this "asymmetric warfare." Others have described the new style of war as "fourth-generation warfare." In this formulation, first-generation war was characterized by close-order engagements with artillery and massed infantry. Large-scale industrial mobilization, enormous destructive firepower, and tremendous losses defined second-generation war. Third-generation war was all about disruptive tactics, like guerrilla war and sabotage, rather than frontal assaults. Today, fourth-generation war often bypasses military operations altogether, attacking political and cultural targets, employing terror and other criminal tactics.

In times past, armies met on a field of battle, engaged with bows and arrows and then in close combat until one side emerged victorious. In the world wars, the last major global conflicts, enormous numbers of men, with huge quantities of materiel, faced one another over vast areas. The invasion of Normandy on D-Day, June 6, 1944, involved over 150,000 Allied soldiers and 20,000 aircraft and ships. With the introduction of nuclear weapons and the beginning of the Cold War, major world powers have refrained from such unrestrained violence, instead choosing to fight in smaller, regional conflicts and often through proxies. Happily, the total numbers of dead in wars has been steadily dropping, but the brutality of war has not changed. Those who still suffer and die can take little comfort in statistics.

In Korea, we fought to prevent the communist takeover of the southern part of the peninsula in something like classic force-on-force battles. In Vietnam we again intervened to "stop the spread of communism" and tried to employ classic tactics and equipment, while the North Vietnamese and the Vietcong engaged in a guerrilla war for which we were not prepared. In 1983, the United States invaded the small Caribbean island nation of Grenada with air assault and special operations forces, to replace a revolutionary government with one more acceptable to us. In 1989, the United States invaded the tiny Republic of Panama to "ensure democracy and human rights" and because its strongman president had "declared war" on us. In the 1991 Gulf War, a U.S.-led multinational force of troops, aircraft, and heavy armor intervened in Kuwait to expel an invasion by Saddam Hussein's forces. In 1992, the United States, operating under the United Nations' auspices, intervened in Somalia, using "all necessary means to establish as soon as possible a secure environment for humanitarian relief operations." In Bosnia in 1995 and again in Kosovo in 1999, the United States and its NATO allies intervened to prevent genocide, but limited their application of force to the employment of airpower. Partly due to the realities of terrain and logistics, the United States employed special operations forces and heavy airstrikes in Afghanistan to find and destroy Taliban camps. In the 2003 invasion of Iraq, the United States used massive airstrikes and again employed troops and armor in its lightning-quick capture of Baghdad. Unfortunately, we didn't plan for what we would do after we won the opening battle.

Now we are faced with so-called hybrid wars or "gray zone" conflicts that are not formal wars and bear little resemblance to traditional "conventional" conflicts between states. As defined by the U.S. Special Operations Command, gray zone security conflicts are characterized by "ambiguity about the nature of the conflict, opacity of the parties involved, or uncertainty about policy and legal frameworks." They do not fit into our well-developed models of war and peace. These conflicts, of which the Russian annexation of Crimea, the rise of the Islamic State, and Boko Haram's terror campaign in Nigeria are examples, combine subversion, destabilizing social media influence, disruptive cyberattacks, and anonymous participants, rather than recognizable armed forces. Future threats will be similarly complex and will require complex responses.

TECHNOLOGY AND WEAPONS

Humans have always found untold reasons and all sorts of ways to kill their fellow human beings. Technology has made that quest even easier. The technological explosion of the nineteenth century was driven in large part by wars. The U.S. Civil War saw the development of a government-funded national armaments industry that continued after the war's end. American industry, science, and applied technology in turn owe a great deal to innovations pioneered by the arms industry.

Especially in the twentieth and twenty-first centuries, developed nations have been enamored of the advantages brought by technology. Societal advances have often followed technological innovation, like electricity, and the needs of society have driven technological development, such as in agriculture. The same is true of the military and its weapons. Nuclear weapons, stealth, technology, and satellite navigation fundamentally and permanently changed the nature of war. Add to this the fact that technologies for both civilian purposes and war are now growing more complicated and interrelated, and it is clear that the outcome of technological developments becomes even more unpredictable.

Wars have changed over time in large measure due to the technologies used to fight them, and the changes those technologies demanded of the military organizations that used them. With the advent of the longbow, Edward III not only introduced a new technology, but modified his organization and tactics as well. The introduction of the machine gun on the battlefield dramatically changed military defensive strategies. And of course, nuclear weapons created almost an entire new military. Separate military cultures developed to protect the weapons and to ensure that they would work. For a portion of the armed services, war was no longer about fighting and winning but rather about threatening yet avoiding the use of those monstrous weapons. The nuclear-capable nations nonetheless felt compelled to pursue nuclear dominance, spending vast sums on more and more powerful arms, and since no country wanted to fight a nuclear war, they continued the buildup of conventional weapons as well.

From the earliest days, advancements in tools and materials enhanced weapons and allowed armies to prevail. The development of steel, for instance, was a watershed event in the history of war, enabling enormous innovation in the design and manufacturing first of knives and swords, then of a great range of other weapons. Rulers and military leaders have supported the advancement of science and engineering. Kings and queens of past eras sponsored royal societies of science, and many technologies, such as civil engineering for fortifications and the science of ballistics, were pursued for military goals. Some of the first presentations to the Royal Society of England dealt with the science of gunpowder, and in the seventeenth century, King Charles I even maintained his own ordnance laboratory in London. In the present day, U.S. presidents employ teams of scientists to give advice on various military technologies.

Countries with vibrant scientific and technological enterprises have been militarily successful, and aspiring countries seek to emulate their approach. Witness the recent phenomenal rise in China's science and technology and its parallel advancements in advanced weaponry.

IMPORTANT WARFIGHTING TECHNOLOGIES, PAST AND PRESENT

The development of weapons follows a path similar to what evolutionary biologists call "punctuated equilibrium," a concept of human evolution introduced by Stephen Jay Gould and Niles Eldredge in 1972. In the biologists' formulation, species evolve slowly over a long period up to a point of sudden rapid development, followed again by a long period of slow growth. Similarly, once a weapons technology is introduced, it is improved incrementally over a long period, then suddenly another technology comes along, adding significant new capabilities and fundamentally changing the battlefield. The so-called punctuations are decidedly nonlinear. They are leaps in capability—discontinuities in an otherwise orderly progression.

Humans first fought with sticks and stones, then swords and knives, arrows and spears, improving them over millennia. The introduction of gunpowder and explosives represented a dramatic jump in the lethality of warfare—a discontinuity. Militaries then improved those technologies for centuries, creating bullets, bombs, and cannons. Machine guns and modern artillery may differ greatly in killing power from earlier rifles and cannons, but they are merely improvements to the same technology.

In the modern era, there have been several notable discontinuous increases in warfighting capability. The introduction of the submarine was an enormous discontinuity in war at sea. While underwater vehicles had been around in rudimentary form for many years, the German development and deployment of the U-boat dramatically changed the course of World War I and subsequent conflicts. The conditions under which submariners had to operate, given the technology of the time, were severe and often dangerous. In 2005, I experienced firsthand the advanced tracking technologies and comparatively comfortable accommodations of a modern submarine on a cruise aboard the USS *Montpelier,* a nuclear-powered *Los Angeles*–class fast-attack sub. These engineering marvels incorporate stealth, high-powered computers, and deadly accurate torpedoes. Other modern submarines are capable of launching intercontinental ballistic missiles. The increase in capability from the early submarines is stunning, but in the end they are still submarines.

Over centuries, armies perfected their tactics and improved their mobility with the use of horses, both for logistics and for combat. The introduction of the first armored vehicles on the battlefield was a discontinuity in capability. Military planners at first struggled to figure out how they would be incorporated with mounted infantry and what purpose they would serve. Since then, their utility on the battlefield has repeatedly been proven. The massive tank battles led by General George S. Patton in World War II and the decisive armored engagements in Israeli wars are a testament to the importance of tank warfare. A hundred years after World War I, I was able to drive an M1 Abrams tank at forty-five miles per hour down a bumpy road at Fort Hood, Texas, and fire the main gun, scoring a direct hit. Aiming and stabilization technologies for lethality and the armor protection afforded by today's armored vehicle are orders of magnitude better than they used to be, but in the end the vehicle is still a tank.

The German development of the V-1, or so-called buzz bomb, and the V-2 rocket in World War II terrorized England. The V-1 presaged the development and introduction of cruise missiles, while the V-2 rocket and its designer, Wernher von Braun, were hugely important in the U.S. space program and the development of ballistic missiles. The V-2 could carry a 2,000-pound warhead. Launched straight up, it could achieve an altitude of 128 miles. NASA is presently developing a rocket that will be more than thirty stories high and will be able to lift 130 tons into low earth orbit.

The British invention of radar, and the U.S. development of it, constituted a great leap in technology that fundamentally changed warfare as well as civil and commercial operations to this day. It is a long way from the rudimentary radars so critical to the Battle of Britain to the MIT Lincoln Laboratory "Haystack" radar, which can image a satellite in geosynchronous orbit, but the basic technology is the same.

The civil space program, so exciting to the public in the 1960s, was in fact a cover for the highly classified development of spy satellites to monitor the Soviet Union's ballistic missile program. Ballistic missiles—themselves enabled by advances in guidance and propulsion—and materials technologies, defined the Cold War era. And of course, the major technological development in warfighting of the twentieth century was the development of nuclear and thermonuclear weapons. The engineering ingenuity, not to mention the national treasure, poured into the design and construction of these weapons is incredible. From the bombs dropped on Japan to the modern nuclear weapon, we've increased destructive power by a factor of a thousand, but they are all still nuclear and thermonuclear reactions.

War today could not be fought without satellites. There are now classified imagery and signals satellites, radar and communications satellites, weather monitoring and global positioning satellites. Military command and control systems depend very heavily on satellite communications, both commercial and military, both encrypted and unencrypted, with 80 percent of the military's satellite communications needs now met by commercial systems. Worldwide monitoring of missile launches, both tactical and intercontinental, is accomplished by sensitive infrared sensors on board high-orbiting systems. They can surveil the entire earth, detect a missile launch, calculate a missile's trajectory, and predict a possible target. They can even make accurate assumptions about the kind of missile based on the heat signature.

Many weapons, not to mention personal electronic devices, could not operate effectively in the absence of global positioning and timing signals. What has become for most people as commonplace and as taken for granted as electricity started out as a system to aid ships at sea. Global satellite positioning systems were a discontinuous leap in the ability of military forces to navigate.

Night-vision systems fundamentally changed the way wars could be fought. No longer were military units confined to daylight operations. While bulky rudimentary infrared imaging was available during World War II, current systems enable a full range of military operations at night and in bad weather. Military pilots are now able to fly at very low levels under conditions of darkness.

Synthetic aperture radars, whose actual motion is critical to the mathematical reconstruction of an image, have demonstrated the ability to "see" through clouds and dust storms. The distance the device travels over a target in the time it takes for the radar pulses to return to the antenna creates the large "synthetic" antenna aperture. A radar wave can penetrate things that would be opaque to an optical camera. Advanced tactical missile seeker technology, which includes radar and infrared, has enabled highly lethal air-to-air, ground-to-air, and air-to-ground missiles to find their targets.

Submarines, tanks, radar, missiles, nuclear weapons, stealth technology, drones, precision-guided munitions, night-vision systems, satellites, and the Internet all dramatically changed the way wars are fought. Each represents a dramatic change in capability that had improved linearly over many years, awaiting the next important development. The point is that, as in evolutionary biology, weapons development and adoption proceeds for many years in an orderly fashion, with constant improvements to a technology driven by external stimuli like a threat or battlefield conditions. At some point, continued improvements become exceedingly costly and reach a point of diminishing returns. Then, suddenly, an innovation or mutation causes a discontinuous leap in capability. With the emerging technologies discussed in the previous chapter, it appears we are now in the midst of a discontinuity.

TECHNOLOGY SEDUCTION AND ADDICTION

In his book *Technopoly,* media theorist and cultural critic Neil Postman writes of the deification of technology, meaning that the culture seeks its authority, finds its satisfactions, and takes its orders from technology. Technology seduces us. Furthermore, once seduced, we become addicted. The analogy with the more traditional context of seduction is a good one, as many of the same processes and incentives seem to be at work, with connotations of persuading someone to do something unwise or overcoming someone's better judgment. The physicist Robert Oppenheimer, in his hearings before the U.S. Atomic Energy Commission, described his seduction thus:

"When you see something that is 'technically sweet,' you go ahead and do it and you argue about what to do about it only after you have had your technical success. That is the way it was with the atomic bomb."

As long as we have developed and incorporated new weapons, we have been focused on the next "sweet" thing or bright, shiny new object. It seems that no sooner do we field one new airplane or ship than we are seeking funds for a newer one. Sometimes new systems are needed to respond to real threats. Sometimes we think we need a new system because the enemy might have one. At other times, however, we see a new capability and just have to have it. I saw this in the development of new satellites. What starts out as a reasonable design grows and grows, because of so many desired new features, until the project becomes unworkable, unnecessarily expensive, or even useless upon arrival. The way we're headed with advanced artificial intelligence, autonomy, and cyber-physical systems seems like more technology seduction.

Technology always promises something better, often with an illusion of objectivity. Solving problems seems to require little subjective thought. If we want more performance, we just need more technology. But technology tends to limit our need to think about alternatives. Unlike problems that must be solved by making a change to an institution or a process, which have multiple competing solutions and may take a long time to show results, technology solutions most often promise immediate impact. That is part of their allure. The demand for more communications bandwidth is an excellent case in point. Rather than modify or cut back on the information we desire to transmit, we merely demand more transmitting capability. In this case, as so often happens, technology provides instant gratification.

One of the U.S. military's top combat commanders, General James Mattis, the former commander of U.S. Central Command, speaks often about the seductive but fragile nature of sophisticated high-tech weaponry and communications systems. He warns that war is a messy, unpredictable affair, conducted by humans, and that technological systems can fail. He concludes that military leaders at all levels need to be prepared for such failures. It was only recently that the U.S. Naval Academy reinstituted, for the first time in ten years, the requirement that midshipmen learn to use the classic sextant for navigation. Can pilots still drop bombs without GPS guidance? Can soldiers still read maps? General Mattis pointed out that in combat, systems fail, and soldiers, sailors, and airmen need to be flexible and innovative to survive without them and complete their mission.

Technologies are addictive. The military's traditional embrace of technological advances has led to excessive dependence on high-tech weapons, as evidenced by the previously described "offsets" and the chronically enormous budgets for military research and development. This willingness to be seduced by technology and our addiction to it are worrisome. Shortages in recent years of so-called rare earth metals are a great case in point. Enamored by ever-smaller and more capable personal electronics and by the promise of wind power for electricity generation, both of which depend critically on rare earth metals, the world panicked when China, which holds the vast majority of global reserves of these materials, restricted their export. More worrisome from a weapons standpoint, rare earth metals are ubiquitous in high-performance aircraft, missiles, and advanced electronics. Without the powerful magnets they make possible, some weapons simply will not work. Ten times stronger than a typical iron magnet, rare earth magnets enable control of the fins on highly maneuverable and very-high-speed missiles. The technology and culture writer Nicholas Carr has said that "the deeper a technology is woven into the patterns of everyday life, the less choice we have about whether and how we use that technology."

Sometimes technical "advancements" make little practical sense and our dependence on them is not much more than wishful thinking. For example, many consider ballistic missile defense technologies essential, thereby justifying the enormous resources spent on them, notwithstanding their somewhat inconsistent operational performance.

Many innovations—computers, satellites, lasers—are now thought to be indispensable. While technology has allowed us to make breathtaking advances in numerous areas, it is our strong dependence on them that makes the users of these technologies vulnerable to their interruption. Satellites are a good example. When one of them fails, the ramifications for commerce and potentially the military can be enormous. The random failure of a communications satellite in the 1990s eliminated pager service and credit card transactions for days until the issue was resolved, with some fairly serious financial implications. Such a failure of critical military systems could have fatal consequences. The breakdown of a satellite link to troops in contact with the enemy could make the difference in their ability to call for air or artillery support. The loss of GPS satellite navigation signals to a precision-guided bomb could cause it to fall too close to friendly troops, as occurred in Afghanistan in December 2001.

Once the recipient of a new technological development sees what sometimes appears to be magical in nature, it is difficult or impossible to take the technology away from him. Battlefield commanders in Iraq and Afghanistan are loath to go on a mission without the latest in satellite or unmanned aerial vehicle (UAV) imagery and are especially intent on ensuring adequate GPS coverage during their missions. In 2002, a senior scientist on assignment to one of the U.S. intelligence agencies from an allied country figured out a way of determining from satellite data if there had been previous, possibly enemy, movement in battlefield areas. On duty one day in the twenty-four-hour operations center where satellite data was analyzed, he detected what looked like a possible ambush in Afghanistan and began sending urgent messages to commanders of the deployed forces. His innovative analysis technique literally saved the lives of an allied patrol and has since become standard practice.

As in the classic sense of seduction, the one being seduced often overlooks or disregards the potential downsides of his actions. The defoliant Agent Orange used so heavily in Vietnam helped U.S. forces uncover enemy hideouts, but created an environmental and humanitarian catastrophe that continues to this day, and that has taken prematurely the lives of many who fought in that war. Land mines and cluster munitions, strewn in huge numbers over various battlefields, have killed or maimed countless innocent civilians. The development of heat-seeking shoulder-fired surface-to-air missiles, commonly known as MANPADS, might have been expedient at the time, and may have been considered essential for tactical success on the battlefield, but their proliferation has created an enormous danger today as they have fallen into the hands of terrorists and other irrational actors. Who can overlook the looming environmental nightmare in the nation's former nuclear weapons facilities such as Hanford, Washington, where millions of gallons of highly radioactive waste sit stored in aging, rusting, and failing tanks?

Compounding the problem of managing or controlling technology, if indeed that is possible, is that predicting how a technology will be used is fraught with uncertainty. One could not have imagined at the outset the myriad uses, both good and malevolent, to which the laser and the jet engine would ultimately be put. Who would have dreamed of a massive Boeing 747 aircraft, carrying tons of hypergolic chemicals (which are highly reactive, releasing enormous energy when they come into contact with one another), with a high-powered laser in its nose capable of destroying a ballistic missile at long range?

Regulators and managers cannot know the full range of the effects of a technology until it has been in use for a while. However, once a technology is deployed and becomes entrenched in society, it is difficult to change or control. Just imagine trying to tell the public it could no longer have smartphones or streaming video. Try to take night-vision devices away from soldiers today.

Technology advances come so rapidly and are in such demand that there is little opportunity to properly consider their potential downsides. Antibiotics are miracle drugs, but their overuse and the resulting resistance of microbes and bacteria has caused the deaths of approximately fifty thousand people in Europe and the United States each year. Chemicals enhance the food supply, but have long-delayed and potentially lethal side effects. The Internet enriches our lives in a multitude of ways, but threatens our ability to think. British neuroscientist and policy adviser Dame Susan Greenfield asserts that the use of electronic devices has an impact on the microcellular structure and complex biochemistry of our brains, which in turn affects our personalities. In short, the modern world could well be altering our very human thought process. This has enormous implications for future combat leaders, not the least of which is the potential for unknown changes in how they will process complex problems and make rapid decisions.

THE ARMS INDUSTRY AND WEAPONS PROLIFERATION

Weapons technologies make it into the hands of warfighters in a couple of ways. In the first instance, warfighters want to mitigate an operational shortcoming and will demand solutions from scientists and weapons developers. Air-to-air missiles are an excellent example. Pilots may learn, through intelligence or actual operations, that enemy missiles have become capable of detecting a target and engaging at ranges longer than our own, thus putting our pilots and aircraft at risk. In electronic warfare, operators want protection against communications jamming, radar spoofing, or intrusion into aircraft identification-friend-or-foe (IFF) systems. In the electromagnetic spectrum, militaries are constantly having to develop countermeasures and counter-countermeasures such as frequency hopping, adaptable waveforms, and sophisticated signal filters.

Occasionally, as with stealth technology, developers may happen upon technologies in the laboratory that could benefit warfighters and will push the technology into their hands. While we had long known of the benefits of properly shaping a structure to deflect radar, it was only the advent of advanced computers and the complex electromagnetic calculations they can perform, combined with materials science and highly precise manufacturing techniques, that allowed us to create stealth aircraft.

Arms manufacturing and export have long served economic and geopolitical aims. A major element of the national security infrastructure of many nations is their capability to produce advanced weapons. The United States spends astronomical sums on weapons programs, more, in fact, than the next seven countries combined. With a total annual defense budget of approximately $600 billion, the United States spends close to $200 billion in research, development, testing, and procurement of new systems. Weapons provide the taxpayer with some vague feeling of "security," but unfortunately they have no real residual value. They certainly can't be repurposed, and most of these systems end up after a few years as scrap metal. Aerial photographs of the aircraft "boneyard" at Davis-Monthan Air Force Base in Arizona reveal acres upon acres of old aircraft, some deteriorating, some wrapped like cocoons, but all operationally useless. Without question, we need a strong defense and a well-equipped military, but we should not conflate the concept of national defense and its often-overused sibling, national security, with a never-ending cycle of more, larger, and higher-tech weapons merely because they are available.

Having spent much of my military career in the business of weapons development and procurement, I have seen firsthand how technologies are introduced, find their way into fielded systems, and are employed. Along the way, I've seen the best of the system, when weapons intended to answer true warfighter needs were produced and fielded in record time and prevented soldiers from dying. Indeed, this was always a top priority, and rightfully so. The soldiers our leaders send into combat deserve the best the nation has to offer. The deployment of systems to counter roadside bombs in Iraq and Afghanistan is a great example of the system working well. The military services' research organizations worked frantically to find ways to detect and defuse roadside bombs. The Mine-Resistant Ambush Protected vehicle, or MRAP, was rapidly developed and deployed and was shown to have saved hundreds of lives in Iraq. I have also seen the worst of the system, when technologies were developed unnecessarily, often for financial rather than operational reasons. Take, for example, the Army's ill-conceived and hugely expensive Future Combat System, on which billions of dollars were spent with little to show for it. Large defense contractors, needing to maintain a certain level of revenue to satisfy shareholders, frequently will take exceptionally optimistic views of the risk of technology developments to win a large government contract. All too often they are wrong. Sometimes weapons, like the Air Force's proposed new nuclear-armed long-range stand-off missile, or LRSO, are developed with inadequate consideration of the potential for unintended or long-term consequences. The LRSO seems to be an unnecessary leap in a pointless nuclear arms race.

Worse yet are cases of shoddy workmanship that injure or kill soldiers. During the Civil War, J. P. Morgan bought defective rifles—soldiers who used them shot off their thumbs—and sold them to generals in the field for obscene profit. More recently, the Defense Department inspector general found that improper grounding or faulty equipment in showers installed by a military contractor in Iraq caused soldiers' deaths. The report concluded that multiple systems and organizations had failed the soldiers.

While the weapons the military develops and buys have changed, the importance of the military-industrial complex has not. The goal is the acquisition of the best weapons possible for our soldiers, but the weapons-buying process has many interested stakeholders, who sometimes forget the primacy of that goal. Industry gets profits, members of Congress get jobs in their districts, universities get research support, and the military gets continuously updated capabilities. A 2014 study by Bloomberg showed that stocks of the four largest Pentagon contractors rose 19 percent in that year, outstripping the 2.2 percent gain for the Standard & Poor's 500 Industrials Index. Weapons purchases create jobs. All but four U.S. states have economic ties to the F-35 fighter aircraft, with eighteen states counting on the project for $100 million or more in economic activity. The project is reportedly responsible for over thirty thousand jobs nationwide. Universities, too, benefit from defense research. The Defense Department spends approximately $3 billion per year on basic research, the bulk of which goes to universities.

Defense- and intelligence-related industries are deeply embedded in government organizations, defense companies surround military research and development centers, and industry plays a large role in helping the government set the research agenda through lobbying and testimony to Congress and DOD executives. Defense contractors provide technical data on systems for which government managers need to write requirements documents and contracts. Many government program offices are staffed by relatively few government employees, often with a third or more being support contractors. Amid an intense competition for defense budget money, weapons manufacturers and technology companies are forced to promise a lot to differentiate themselves from their competition. The F-35 fighter competition promised to net the winner a $200 billion contract. Both bidders promised vertical takeoff and landing (VTOL) and other advanced capabilities. Lockheed Martin's design won over Boeing's, but the VTOL capability came at great cost in money and time. To date, the fighter is still short of meeting all of its requirements.

Industry primarily, but government organizations as well, has a vested interest in a continuous flow of funding and requirements for new technology and weapons to assure their own organizational survival. Given the massive amounts of money at stake, we need to be sure that we understand the motivations and pressures surrounding weapons procurement. We can at least begin to do that by questioning and being more skeptical of the relentless increase in requirements for new weapons. Essayist James Fallows, writing in *The Atlantic,* noted that "we buy weapons that have less to do with battlefield realities than with our unending faith that advanced technology will ensure victory, and with the economic interests and political influence of contractors."

Not only are we addicted to a constant stream of new weapons for ourselves, but we are keen to export our sophisticated weaponry. We are huge weapons proliferators, selling more weapons to more countries than any other nation in the world. In 2014, the United States sold more than $10 billion worth of arms to other countries, while Russia sold almost $6 billion. The two countries account for over half of the global market. Customers for U.S. arms include Saudi Arabia, the United Arab Emirates, Turkey, Vietnam, Iraq, Egypt, and Pakistan. While we restrict the sales of our most advanced systems, or limit them to our closest allies, previous versions of "old" technology remain no less deadly.

We lead the world when it comes to weapons research and development, and as leaders, we also have a responsibility to demonstrate caution as an example to others. In 2012, then national security adviser John Brennan, speaking about the use of drones in the war on terror and the fact that many other nations were seeking the technology, said that "if we want other nations to use these technologies responsibly, we must use them responsibly," and that "we cannot expect of others what we will not do ourselves." But what constitutes responsible use? If the United States uses a weapon, like an atomic bomb or an armed drone or napalm, it is ridiculous for us to think we can prohibit others from doing so.

If countries are going to persist in proliferating high-tech weapons around the globe, then they have the responsibility to also provide training in the proper use of those weapons. That does not always happen. Quality training is time-consuming and costly. It is widely believed that the surface-to-air missile that brought down the Malaysia Airlines Flight 67 civilian airliner over Ukraine in 2013 was a Buk antiaircraft system supplied by Russia. These are highly sophisticated systems that could have easily distinguished between a military and a commercial aircraft. I continue to believe the tragedy occurred because the operators of the system were ill-trained by the providers.

Governments have engaged in important debates about nuclear arms reduction treaties, conventions banning weapons of mass destruction, and weapons that result in unnecessary and superfluous suffering, such as land mines. Dual-use technologies like artificial intelligence or synthetic biology present a different dilemma in that they have the potential to be used in both good and evil ways. While the technologies themselves are not the subject of treaties and conventions, we are now faced with controlling the proliferation of weapons employing these technologies.

WHERE WE'RE HEADED

What the future has in common with the present and the past is that the United States has played and will continue to play a leading role in science, technology, and innovation. This has been true in both the civil and military domains. Today, however, many of the complex new technologies are available to other advanced nations, friends as well as enemies. While in the past, weapons technologies were specific to military uses, today's important technologies are increasingly dual-use. Many have decidedly important civil applications.

Whereas weapons technology once tended to improve range, speed, lethality—tactile attributes—new technologies, such as autonomy, soldier enhancement, and artificial intelligence are fundamentally altering the meaning of conflict and soldiering. The demands placed on the individual soldier are very different today. Past weapons were tools placed at the disposal of the soldier. Largely improvements on previous weapons, they had more power, speed, or accuracy. In addition to being far simpler to operate technically, they required little thought beyond the tactical and military expertise for which the soldier was trained.

Today's weapons demand an understanding not only of the technology but of far more complex rules of engagement. They do indeed require greater knowledge and care, and the soldier's interaction with the enemy changes in fundamental ways. Former DOD official and now Georgetown University professor Rosa Brooks contends that the line between war and not-war is blurring in part because of advancing technologies. "If we get to a point," she says, "where we can no longer tell the difference, that could fundamentally challenge the law of armed conflict."

Every era of war and weapons has come with its own set of ethical issues. The introduction of the machine gun into combat was met with horror; nuclear weapons have almost universally been deemed morally wrong; and biological and chemical weapons have been outlawed as inhumane (which doesn't mean they will not be used). Are we going to place these new technologies and systems into the hands of our soldiers and ask them to figure out when, why, and how to use them? Are we going to ask our military to figure out the appropriate employment of such weapons? Policy regarding these systems is lacking. Training is lacking. We cannot expect young soldiers to understand the nuances. There are risks, both technical and otherwise, in deploying technologies that we don't completely understand. No one anticipated the psychological toll that drone strikes would have on their pilots. We cannot possibly anticipate how an autonomous system will react to a confusing and ambiguous situation. How will the presence of hypersonic weapons, against which there are currently no effective defenses, change the behavior of an adversary? We must urgently consider how these developments will ultimately affect not only the enemy, but also the soldiers who are asked to adapt to them.

It is understandable that militaries would retain a research infrastructure to work on future ideas. New types of warfare call for different types of weapons. What is troubling, however, is the utter fascination we have with applying new technologies to our weapons, with far too little questioning and skepticism.

We know that technology has had unintended and dire consequences. No one could have foreseen the damaging environmental results of carbon emissions from internal combustion engines, or the reckless nuclear arms race resulting from the discovery of nuclear energy, or the fundamental ways that computers and advanced communications have changed human existence. None of this is to say that we shouldn't have pursued these technologies or even the resulting systems. That would be naïve and counterproductive. It does say that we have to work much harder to anticipate the consequences of our actions. These new tools bring new ways of inflicting harm and also new challenges to the laws of armed conflict and warrior ethics. It is the responsibility of scientists to think deeply about the potential consequences of their work, and it is equally important that the public, the media, and national leaders explicitly discuss and debate those implications early, before technologies are widely deployed. Coming weapons technologies will demand more understanding, more training, and greater debate before they are employed. These should not be left to our young soldiers to figure out on their own.

3

EFFECTS OF FUTURE WAR ON THE SOLDIER

A YOUNG SOLDIER *is leading his small combat team as it makes a forced entry into enemy territory. Alongside him are a few of his buddies from basic training and several robots, some with human physical characteristics and others designed with unique features like tracks for mobility over rough terrain. The soldier's energy and, frankly, his enthusiasm for the mission are flagging when he receives an order in his combat helmet to inject a drug for energy and mood alteration. The team comes upon a structure and the robot leader's wall-penetrating radar indicates the presence inside of potential hostiles. Taking up combat positions, the team prepares an assault on the house on orders from the commander, who is watching the operation on video from a rear area. Our young soldier is not convinced, noting the absence of any other indications of enemy presence. He relays his concerns privately to the lead robot, who he knows is recording the exchange, but he is reluctant to abort the mission, thinking the machine has more and better sensors than he does. The rear-area commander, impatient with the delay, orders the lead robot to clear the house with all necessary force. When the dust settles, the team finds the* *bodies of a peasant farmer and his family with pitchforks and clubs, huddled around an old radio. The young soldier, without any orders, injects a drug to forget.*

Soldiers, most of whom are young, experience unimaginable horrors civilians cannot fathom.

Not long ago, Major William Martin, chaplain, U.S. Army, discussed with me his experiences during one of his several deployments to Iraq. Sixteen soldiers in his unit had been killed and over a hundred wounded. Lacking a strong commander, the surviving soldiers felt despondent, angry, and helpless. He described his daily challenges of offering hope and pastoral care to scores of soldiers "anguishing from the effects of war and the wounded soul." It is Martin's strongly held belief that the teaching of the concepts of Just War Theory, with its recognition of the legitimacy of war and its emphasis on proper behavior and its moral basis, provides soldiers with an emotional structure that serves to soften or mitigate the damage war does to their souls. If justice and morality as described in Just War Theory were part of combat training, he believes, moral injuries could be better addressed and treated. Martin's view is that morally neutral modes of being that ignore ethics and couch combat operations in more utilitarian terms exacerbate moral injuries and frustrate their treatment. If the chaplain's soldiers had had a leader who helped them understand clearly that war is not merely murder and mayhem, and that it is governed by a set of rules, that their actions in war were acceptable, the harm to their psyches could have been less severe.

Once in combat, the soldier has life-changing encounters. Author and activist Peter Marin has said that "the mistakes one makes are often transmuted directly into others' pain; there is sometimes no way to undo that pain—the dead remain dead, the maimed are forever maimed, and there is no way to deny one's responsibility or culpability, for those mistakes are written, forever and as if in fire, in others' flesh." Soldiers need to be trained emotionally as well as physically, and they need to be monitored once their tour of duty is over.

We ask an enormous amount from our military men and women. Repeatedly we send them into battle to settle matters of state or to defend those who cannot defend themselves. Our soldiers, and those of other nations, are given special authority by the state to engage in lethal operations, without which they would be considered murderers. General John Allen, former deputy commander of U.S. Central Command, says that it is the intention of the warrior to defend not his or her own life, but the lives of others, that constitutes the individual combatant's moral justification for waging war and exempts them from blame for harming or killing others. The Just War philosopher Michael Walzer refers to this as "the war convention."

Given such awesome responsibilities—to kill on behalf of the nation and its citizens—soldiers understandably have a sense of specialness. Soldiers tend to view themselves, rightfully, as members of a unique segment of society. They are charged with guarding a sacred trust to protect and honor the values of the nation. As General Douglas MacArthur said, "The soldier, be he friend or foe, is charged with the protection of the weak and unarmed. If he violates this sacred trust, he profanes his entire culture."

Because of the unique and highly stressful nature of their position, soldiers create for themselves a special fraternity, with its own rules, rites, and behaviors, including solemn honor guards, deeply touching burials featuring a riderless horse, the solemn playing of "Taps," and gun salutes. Soldiers even humorously touch on death in songs and ditties like "Blood upon the Risers."

Killing is naturally abhorrent to most humans. Obviously, as I said, soldiers are faced with horrors most people cannot imagine, let alone ever see. I spoke with an army officer from an African nation where child insurgent soldiers were common. Clearly distraught, he reported that his army was suffering excessive casualties because its soldiers were hesitating to kill the child attackers. The child soldiers, many operating under the effects of drugs they were given, had no such compunctions. In the end, the army's rules of engagement called for them, sadly but correctly, to treat the children as enemy combatants. General Allen has noted that "unchecked human savagery in war delivers our young troops to the edge of an abyss where indescribable physical and psychological scars are inflicted."

Combat-related psychological problems have been with us throughout the history of war. What we now call post-traumatic stress disorder (PTSD) has previously been called battle fatigue, Vietnam syndrome, or shell shock. In a discussion of the distinction between PTSD and the more moral injuries of the type discussed by Major Martin, author Maggie Puniewska relates the stories of soldiers in Iraq and Afghanistan who violated their own moral code by hurting an unarmed civilian, or shooting at an armed child, or not acting on something they witnessed and should have tried to prevent. She concludes that moral injury is not about loss of safety, but loss of trust—in oneself, in others, in the military, and sometimes in the nation as a whole. When our soldiers have a moral basis for their actions, they have a scaffolding, or support system, on which to lean. When they don't, the things they are required to do, to win or merely survive, are so far outside their sense of normal behavior that they may be traumatized.

For those whose idea of war is video games and movies, this may seem like so much talk. But the soldier in the field is not a Hollywood actor—he sees things we will never see, experiences things we will never experience. It is comforting for him to be able to draw upon centuries of experience and rational thinking about war.

The images of war are grotesque and sad beyond belief. Photographs of the dead in the Civil War or in World War I attest to the carnage and make us contemplate all of the young lives cut short by differences in ideology. Descriptions of the Crusades and the Inquisition and recent videos posted by ISIS show us the brutal and reprehensible behavior by supposedly religious people in the name of God. Images and descriptions of the heinous mass murder of the Jews by Hitler in World War II or of the grotesque biological experiments by the Japanese on prisoners of war and innocent civilians cause us to question the capacity for man's inhumanity to man. The deliberate targeting of civilians by Allied bombers over Germany and Japan and the dropping of the atomic bomb—twice—on civilian targets, or the use of Agent Orange in Vietnam, should cause us to at least think carefully about the ramifications of developing new lethal technologies. In applying to war new technologies developed for other purposes, we must plan for the possibility that they will be used in inhumane ways and specifically describe rules for their employment. In concept, nonlethal weapons are good. In practice, they can be used as means of torture. Neuroscience can be used to enhance a soldier, but can also be used as a tool for interrogation.

THE LAWS OF ARMED CONFLICT

For millennia, the brutality of war and of leaders and tyrants was unrestricted. Early warfare knew few bounds. However, in more recent centuries, thoughtful people have searched for ways to mitigate the misery brought on by war. They have sought to minimize the brutality and wanton destruction that occurs on the battlefield and have developed theories of "Just War" and "laws" of armed conflict. Essentially, they have tried to establish standards for ethical behavior of soldiers and parties to conflict, and ways of protecting those not directly involved.

Just War Theory and the laws of armed conflict are intended to guide militaries in their behavior in war. These standards of behavior influence the actions of our soldiers and contribute to the so-called warrior code. Australian ethicist Robert Sparrow has written that "while war remains a ghastly business, when the standards are maintained, warrior codes function to reduce the horror of war and tame the worst excesses of young men sent out to kill strangers in foreign lands with weapons of terrifying power." Justifiable reasons to go to war and guidelines about how combatants behave are timeless needs. They are as important today, and will be in the complex wars of the future, as they have been for millennia. By definition war is awful, but civilized leaders still see the need to apply limits to it, lest it spiral out of control and into chaos and butchery.

Notre Dame professor Don Howard and I have written that "Just War Theory has ancient historical roots, important parts of which lie in various religious and philosophical traditions, as expressed in works as diverse as the Old Testament Book of Deuteronomy, the Qur'an, and the Indian Mahabharata. In the form in which it later came to shape the international law of armed conflict, however, the theory is mainly derived from the works of early and medieval Christian thinkers including Augustine (354–430) and Aquinas (1225–1274), who were among the first to formulate explicitly such principles as proper authority, just cause, and right intention." The Dutch statesman and scholar Hugo Grotius observed in the seventeenth century that there was a "lack of restraint in relation to war, such as even barbarous races should be ashamed of, that men rushed to arms for slight causes, and that when arms had once been taken up there was no longer any respect for law, divine or human." Aiming to change that, he played a critical role in systematizing the rules of armed conflict. It was in his 1625 book *On the Law of War and Peace* that the modern principles of Just War Theory were first codified and that the fundamental distinction between justice in the decision to go to war *(jus ad bellum)* and justice in the conduct of war *(jus in bello)* was made clear.

Decisions to go to war *(ad bellum)* must be based in rationality, on just causes and right intentions. Wars must be started by someone with the proper authority to do so. War must be proportional and not, for instance, a response to a mere slight by another party. War must be undertaken as a last resort, after all options to avoid it have failed. These were seen as questions to be taken up by the state and involve decisions that are beyond the day-to-day concerns of the individual soldier.

Several key rules dictate proper conduct in war *(in bello)*. Force may be used only against legitimate targets, meaning active combatants. Civilians or other noncombatants should not be targeted under any circumstances. Only appropriate and proportionate levels of force may be applied, and only that force required by the pursuit of a legitimate military objective. No methods or weapons may be used that are evil in themselves. Future wars, characterized as they will be by uncertainty and ambiguity, by technologies that may be indiscriminate and unpredictable, and that are fought increasingly among the populace, will make these rules of conduct even more important.

The rules of war first found expression in the nineteenth century as explicit instructions for armies and explicit international agreements. The first important handbook for troops in the field was the Lieber Code of 1863, prepared by attorney Francis Lieber at the request of President Abraham Lincoln. Lincoln's War Department was deeply concerned about the maintenance of discipline in the Union army. The Civil War was a particularly brutal one, with excesses on both sides, and atrocities would make maintaining the eventual peace more difficult. Lincoln was interested in *jus post bellum,* or justice after the war. Promulgated as General Orders No. 100, the Lieber Code, distributed to all Union forces, included detailed rules for such areas as military jurisdiction, the protection of property, assassination, insurrection, and the treatment of prisoners of war, deserters, partisans, and spies.

The modern framework for the international law of armed conflict and international humanitarian law was established with the Hague Conventions and the Geneva Conventions. The former established general principles for the resolution of international disputes, the conduct of war, and the rights of neutral parties, along with specific provisions concerning poison gas, expanding bullets, aerial bombing, submarine warfare, and mine laying. The latter established rules for the treatment of the sick and wounded, prisoners, and noncombatants.

The past century has seen numerous attempts to come to grips with and mitigate the effects of war and its technologies. After World War I, an international agreement banning the development and use of chemical weapons was signed. Since World War II, numerous efforts have been made to deal with issues of technology, weapons research, and ethics. These include the 1946 Nuremberg trials, the 1972 Biological Weapons Convention, the 1993 Chemical Weapons Convention, and efforts by scientists to place restrictions on biomedical, genomic, and nanotechnology research. Scientists attending the Asilomar Conference on Recombinant DNA, near Monterey, California, in 1975 recognized the potential dangers of such DNA research and declared a moratorium until safe and ethical procedures could be developed. The guidelines developed were voluntary, but have been assiduously followed.

Rules and theory are one thing, practical applications another. Philosophers have advanced two basic theories that have proven useful in analyzing moral problems and making decisions about ethical issues. One theory, consequentialism, asks which actions will provide the greatest net good for the greatest number of people when considering both harm and benefit. An action is judged to be right or wrong given its consequences. Thus a soldier is asked to consider this balance in determining an acceptable level of collateral damage in military operations. A second theory, called deontological, or Kantian, ethics, judges the morality of an action in terms of a set of strict rules concerning our duties, rights, and justice. The morality of an act, such as the decision of whether to use torture, is to be based on the nature of the act and not on its results or on the identity of the actor.

A hybrid philosophical approach to decision making is called virtue ethics. It emphasizes good personal character as most basic to morality. The core values of each military service embody a virtue ethics approach. The individual soldier is often placed in a difficult position, fighting in what he may consider an unjust, or at least unnecessary, war, but attempting to fight in accordance with proper rules of conduct.

In the years since the Vietnam War and the ensuing professional growth of the military, much has been made of the warrior ethos in an attempt to genuinely mold the military into a professional service. Ethos for warriors, as with other professions, describes a way of thinking and acting that incorporates traditions, strictures, and professional goals: the Spartans are often held up as exemplars of such an ethos, with their tightly controlled society and intense martial spirit.

It was the concepts of Just War Theory developed by early theologians and the related concepts of chivalry extant in the medieval period that explain much of the warrior ethos as it exists today. The medieval code of chivalry included traditions involving bravery, warrior professionalism, and service to others. What eventually became Western warrior customs—the warrior code—began to emerge between the eleventh and fourteenth centuries. Over time, the word "chivalry" came to describe social and moral virtues more generally. The Dutch historian Johan Huizinga described the code of chivalry as a moral system that combined a warrior ethos incorporating bravery and courage with a social mien including demonstrations of honor, knightly piety, fidelity, and courtly manners. Oxford University law professor Theodor Meron wrote that "the humane and noble ideals of chivalry included justice, loyalty, courage, mercy, the obligations not to kill or otherwise take advantage of the vanquished enemy, and to keep one's word."

These may seem like quaint notions in the current age, but as Professor Meron noted, "the idea that chivalry requires soldiers to act in a civilized manner is one of its most enduring legacies." The ability of soldiers on battlefields of the future to understand, still more be able to apply, such notions will be an important challenge. As we will see, advanced weapons and new styles of war will change the relationships of soldiers to the enemy and to other soldiers.

THE IMPORTANCE OF LEADERSHIP

No amount of teaching or training can completely prepare a soldier for war and the ugly realities of combat. Instruction is, however, critical in inculcating a way of thinking in the soldier so that when faced with demanding situations, he or she knows instinctively a proper response. In the heat of battle, survival and mission accomplishment naturally take priority. Good, consistent, continuous training will help the soldier in making correct split-second decisions under extreme circumstances. Leadership, however, is crucial, both on and off the battlefield. The command climate established by a military leader has a huge effect on the behavior of deployed troops.

In 1968, in the South Vietnamese village of My Lai, Lieutenant William Calley and his men systematically gunned down over one hundred villagers, including women and children. The case of Lieutenant Calley and the massacre at My Lai, originally reported by investigative reporter Seymour Hersh, is well known, as is the command climate of obfuscation and deceitfulness all the way to the top.

In 2003, U.S. soldiers engaged in torture and murdered prisoners in Iraq, seriously, perhaps permanently, damaging the standing and reputation of U.S. forces there. The entire chain of command, from General Ricardo Sanchez, the commander of the Iraqi task force, to Brigadier General Janis Karpinski, commander of the military police unit running Abu Ghraib prison, near Baghdad, was clueless and uninterested. What must have been the command climate in the unit of U.S. Marines pictured urinating on enemy corpses in Afghanistan in 2012?

World War II general Curtis LeMay, who commanded the B-29 bombers that conducted incendiary bombing raids on Japanese civilians, set a dangerous command climate when he said he believed there were no innocent civilians, that we were fighting the Japanese people. "So it doesn't bother me so much to be killing the so-called innocent bystanders."

General Colin Powell and President George H. W. Bush, however, in the final days of the Gulf War, set a different example. As Iraqi forces were on the run, they decided, in a move controversial to military hawks but that they considered to be ethically correct, to call off the attack and let Iraqi forces retreat to Baghdad.

Finally, consider the real-life story of the Navy SEAL team depicted in the movie *Lone Survivor.* In that case, the commander decided for his team what he felt was the morally correct option to spare the lives of local civilians, with unfortunate and ultimately tragic, deadly consequences for his men.

Overwhelmingly, soldiers are highly professional and intent on succeeding in their assigned missions while upholding the standards and traditions of their service. Leaders place them in terrible positions when they do not lead by their own example. It becomes difficult, if not impossible, for a young soldier to maintain his values when even our national leaders find weak and spurious justifications for abhorrent behavior like torture, as they did during the Iraq War.

Leadership and command are critical, for they set the ethical standards for the unit. Leaders and commanders guide the behavior of their soldiers. The diffuse nature of responsibility inherent in future technological warfare will challenge this concept. The future soldier will be hyperconnected, operating as part of a network with his every action monitored and analyzed by computers. Military units will employ humans and robots and will depend heavily on artificial intelligence systems. It may be difficult to know who is actually in charge.

Leadership and good training and education are key, to be sure, but they are continually challenged by the appearance of new technologies on the battlefield. Weapons development is rarely accompanied by ethical reflection on its consequences. On occasion, developers have sought to couch their work in peace-loving terms, reflecting either self-justification or a dangerous naïveté. Alfred Nobel proposed that his invention of dynamite would have such a momentous effect that future wars would be impossible. Richard Gatling wanted his invention of the machine gun to make war so horrible that it would end wars altogether. It hasn't quite worked out that way.

Sometimes, however, the most horrific of weapons, such as nuclear and thermonuclear bombs, have had the effect of being so awful in their results that the world has disdained their use. The development and rapid rise of precision-guided munitions such as television-guided bombs, laser-guided bombs, and GPS-guided bombs, has had positive ancillary effects. Such munitions, while not perfect, are orders of magnitude more accurate than their predecessors. Bombing of targets once required enormous numbers of bombs and bombing runs, which was expensive and dangerous for pilots, created a great deal of collateral damage, and killed many innocent civilians. Precision-guided munitions can destroy targets with fewer, or perhaps single, bombs. As a result, there is less danger for pilots and less collateral damage, a genuine moral gain.

WHY WORRY ABOUT ETHICS?

The fact is, technology continues to make killing more efficient, and the idea is to win wars. When faced with questions or concerns about ethics, then, the first reaction of technologists, weapons developers, and decision makers is to find reasons why the discussion is unnecessary. Since our adversaries are unethical, why should ethics be a constraint in using our most advanced weaponry against them? Adversaries, it is said, will pursue all technological opportunities that serve their interests, and if we don't do likewise we'll be at a military disadvantage. These arguments are valid, but place pragmatism ahead of moral thinking and virtually dismiss moral or philosophical introspection. What will be the potential consequences of our actions?

One consequence is that once moral restraints are ignored, it is easier to ignore them again. In addition, we may not always fight fanatics, and our use of brutality or questionable weapons will be well noted by others. In the period after 9/11, the United States resorted to torture of enemy detainees. While most senior leaders have denounced the practice, the fact remains that the nation crossed an important moral threshold. Knowing that, future enemies—even civilized ones—may be less inhibited in employing the same methods against us.

The United States leads the world in many areas of technology, including military technology, to be sure. It remains to be seen, of course, how future U.S. leaders will view these issues, but we have historically purported to be strong proponents of human rights. The things we do will, like it or not, drive the behavior of others. If we can engage in targeted killings, how can we complain if others do the same? I have spoken publicly in favor of the agreement with Iran concerning its nuclear program. Like any thoughtful person, I do not desire to see Iran or anyone else build nuclear weapons. But really, if we have nuclear weapons, how can we tell others they can't, and go so far as to talk about starting a war to prevent them from getting them? If we are going to lead, we have to be ethical leaders ourselves.

Certainly our adversaries may be unethical, but it just doesn't follow that we should become the same barbarians we presume our enemies to be. It is not what our adversaries do that should drive our decisions. We should only do what is morally justifiable within our own frameworks. The United States has always prided itself on its values. While such an argument might appear naïve, our behavior is what sets us apart from our enemies. In the face of the brutality of the Taliban and other terrorist groups, General David Petraeus felt the need to remind his soldiers in Afghanistan of the behavior expected of American troops, saying, "While we relentlessly pursue and kill our enemies, we must observe the standards and values that treat noncombatants and detainees with dignity and respect. While we are warriors, we are also human beings."

Researchers and developers are generally conscientious, well-meaning, and sensitive to legality, but they rarely consider ethics. Surprisingly, there is no prescribed ethical review or point during a program at which weapons technology researchers are asked to explicitly consider the implications of their work. Indeed, in my long technology-focused career, I never once heard—or asked—the question of whether we *should* be developing a particular technology. We must ask ourselves: Does the system encourage, or even allow for, ethical reflection or review? When can and should participants question a technology or a weapon, and when should they raise objections?

Some writers take a very dark view of military ethics and the laws of armed conflict. They feel that ethics in war allow us, as Major Ralph Peters has said, "to disguise psychologically the requirement to butcher other human beings, masking the necessary killing behind comforting concepts" such as Just War Theory. Military ethics, Peters says, implies "a comforting order in what is really chaos and void." Much as Major Martin believes that Just War Theory can heal a soldier's wounded soul, Peters's position is that so long as we believe we have behaved ethically, we can psychologically bear the knowledge of our deeds. While this cynical view has a shred of truth to it, it does a disservice to soldiers who attempt to hew to a code of ethical behavior.

The philosopher Michael Walzer says, "Though chivalry is dead and fighting unfree, many professional soldiers remain sensitive (or some of them do) to those limits and restraints that distinguish their life's work from mere butchery."

The laws of armed conflict have been extraordinarily successful in the past because the conflicts were between states, or so-called rational actors. Unfortunately, today's threats often come from failed states and armed groups that look upon the West's adherence to rules of war as ridiculous and capitalize on our ethical stance. The Canadian author and former politician Michael Ignatieff explains that "we in the West start from a universalist ethic based on the ideas of human rights, while our adversaries start from particularist ethics that define the tribe, the nation, or ethnicity as the limit of moral concern." We can look at this as either a weakness that needs to be eliminated, or as a strength of our society. I choose the latter. To do otherwise implies acceptance of barbarism and the rejection of centuries of experience with what constitute civilized behavior and civilized nations.

All we understand about the laws of armed conflict is based on old styles of conflict, pitting soldier against soldier on the field of battle. We recall that the four main principles derived from these theories and the laws of armed conflict are that soldiers must: (1) practice discrimination, doing everything in their power to target only combatants and not the innocent; (2) ensure proportionality in their actions, using only that amount of force necessary and proportional to the situation; (3) consider absolute military necessity, engaging in violence and lethal operations only when the military situation demands it and no other alternative is possible; and (4) minimize unnecessary suffering, using only those weapons and techniques that cause the least amount of suffering commensurate with their task. New weapons and styles of conflict fundamentally change the role of the individual soldier and will demand a new understanding of how to apply these venerable rules.

War is a deeply human activity because the battlefield gives rise to such feelings as courage, fear, cruelty, remorse, altruism, guilt, sacrifice, and empathy. The traditional battlefield required an interaction with people, both friendly and enemy. Soldiers fought in units and depended upon one another for their lives. Trust was critical. On battlefields of the past, killing was intentional and intensely personal. On battlefields of the future, technology will deaden our senses to the horror of war and its consequences.

IMPLICATIONS OF THE TECHNOLOGIES

The increasingly automated nature of combat, the artificial enhancement of soldiers, and the speeds and distances involved threaten to undermine the warrior ethos. New technologies confound long-held rules of war, or at least our understanding of how they are to be applied. We have already discussed some of the advantages and possible practical disadvantages of new weapons and technologies, and the attendant ethical and moral issues.

New styles of war, and thus new technologies, will obviously affect soldiers. How does a particular technology affect what it means to be a warrior? Autonomous weapons, cyber operations, and soldier enhancements change the sphere of battle. Unmanned vehicles and directed-energy weapons, for instance, portend new battlefields and different relationships between combatants. Killing becomes impersonal. The military and societal implications of robots and enhanced soldiers will also be enormous. Not only will these new technologies change the relationships between opposing soldiers, but they will also affect soldiers on the same side, and may alter the meaning of important concepts like trust and altruism. In the past, soldiers took risks for one another. It is not at all clear how the introduction of autonomous machines will change that dynamic, or if a soldier will entrust his life to a machine.

We have all heard heartrending stories of bravery, heroism, and self-sacrifice, such as a soldier giving his life by falling on a grenade to save others in his unit. Imagine a situation in which a robot is faced with a decision as to whether to expose itself to enemy fire to save soldiers pinned down by the enemy. The robot might well do a quick calculation and decide that in that day's battle it has been the most efficient member of the team at destroying the enemy, and that its loss would upset the casualty ratio. So much for loyalty to one's mates. And assuming a robot is doing a great job defending the rest of a unit in combat, will a fellow soldier put his life or the lives of other soldiers in danger to save the robot? Should he?

If we can ever build a functioning machine that can approximate human behavior and demonstrate rudimentary moral decision making, what about the men who fight alongside those machines? How do the two emotionally connect? Camaraderie becomes a meaningless concept. Consider, for example, the frequently imagined future where humans and robots operate in teams on the battlefield. One could envision a situation where the human gives an order, perhaps even a flawed order, with which the robot disagrees. Can the robot refuse? If so, are there consequences? The idea of punishing a robot is ridiculous. Or consider the case where a soldier has to make a questionable decision and only has seconds to do it. Will he be inhibited by the presence of the robot that sees everything and forgets nothing? These situations will have major implications for unit cohesion and trust among soldiers, and trust is crucial in battle.

Soldiers will no longer have to fear close contact and hand-to-hand combat because they will be able to deploy robots and unmanned vehicles at great ranges. This will of course greatly reduce friendly casualties. However, fear acts as a modulator of behavior, and by reducing it we might also remove constraints on unethical behavior. A soldier deciding whether to enter a village no longer has to fear coming under small-arms fire when he has the option to call for artillery or airstrikes, or to send in an armed robot. Now that war is increasingly fought at long distances, soldiers have become too distant from their own actions. This has major consequences. Fighting at long range and seeing the enemy only as a "target" or an icon on the screen dehumanizes him and implies we have no chance to demonstrate emotions such as empathy, important when we're authorized to kill, and perhaps key to decisions about military necessity or proportionality. If a soldier cannot see, hear, and understand the context of a battlefield or a particular engagement, he is less likely to concern himself with decisions requiring such nuance.

In the early days of the armed Predator drones, video footage of the missile strikes was often available on the classified military Internet system. At a stateside Air Force base, I noticed a group of airmen huddled around a computer screen, watching as a truck full of armed Afghan men drove down a mountain road. Suddenly there was an explosive flash of the missile, and truck parts and body parts were sent flying. In the accompanying audio you could hear laughing and cheering, as if the whole thing were some form of entertainment. The film was graphic and gruesome and the surreal scene has stuck with me.

An elderly World War II combat veteran told me in confidence of his still vivid memories of his fear when on patrol near enemy lines and the sheer terror each time he was in combat. He also told me of the awful feeling of having to look someone in the eyes when you stab him to death with a bayonet. If we cannot see the enemy, it is a lot easier to kill him.

Soldiers will have enormous amounts of data, delivered

to them faster than ever and over more media because of better sensors, power, and communications systems. More data, however, does not necessarily mean better information. Greater dependence might well dull the use of common sense and warrior instinct. A soldier might be unwilling to ignore the data, even when his instinct tells him otherwise, or be enslaved to it even when it may result in a morally questionable action. Without a personal dimension to war, how can we hope to inculcate those values that have served us for so long? Human beings make ethical decisions by being a part of society, through the experience of emotions and interaction with others. Sensory experience and social mechanisms—in short, a lifetime of experiences—contribute to our ways of behavior. Former naval intelligence officer William Bray writes that "human analysis also relies heavily on subjective experience, which too often is taken for granted, but is central to understanding anything. Every piece of data perceived is understood only in the context of everything a person has experienced." Machines cannot replicate that.

It is doubtful that we can expect machines to make nuanced decisions concerning the distinction between combatants and noncombatants on the battlefield or the protection of noncombatants from collateral damage. Perhaps a machine can make a utilitarian calculation on proportionality of force, but can it make an empathetic decision on what is unnecessary and superfluous suffering? Will it be able to sense an enemy's wavering determination and call off an attack? Satellites and drones can tell a commander of massed enemy troop formations, but they are unlikely to be able to tell him anything about the fierceness of enemy soldiers or their will to fight. The

United States was rightfully concerned about large numbers of Iraqi Republican Guard troops seen by surveillance sensors to have been massed in the Kuwaiti desert in the Gulf War, and based on that data, U.S. forces were prepared to apply overwhelming force and firepower. But when confronted by U.S. troops, the Iraqis readily surrendered.

Machines, both physical and virtual, will engage in conflict either alongside human soldiers or without them. Autonomous weapons and long-range weapons allow conflicts to be fought from afar, mitigating the need to place soldiers in harm's way. While it is comforting to know that our soldiers will not be hurt, war and violence should not become too easy. What is going to put a check on the propensity to resort to violence if it can be done without any threat to us? Clearly these technologies impact decisions about last resort and military necessity. Decisions by commanders to attack an objective have always been based partly on a calculation of risk, both to our forces and to enemy noncombatants. Risk becomes a lesser factor with autonomous robotic systems.

The demand for immediate action leaves no room for delay. The philosopher Simone Weil said that "the man who is the possessor of force seems to walk through a non-resistant element; in the human substance that surrounds him, nothing has the power to interpose, between the impulse and the act, the tiny interval that is reflection," and that "where there is no room for reflection, there is no room for justice or prudence." Modern weapons leave little room for reflection.

Increasingly, warfighting will exceed the capabilities of the human senses to collect and process data. Biological sensors and human thought processes cannot hope to achieve the exquisiteness and speed of computers and advanced electronics. Computers, artificial intelligence, robots, and autonomous systems will create an environment too complex and fast for humans to keep up with, much less direct. Gradually, perhaps imperceptibly, automated systems will function so much more efficiently than men and women can that we will become bystanders. Future conflicts will increasingly be defined by speed-of-light weapons and computer automation. Human perception and coordination will become a limitation. The soldier will become the slowest element in an engagement, or simply irrelevant. A soldier who feels himself to have no autonomy, authority, or responsibility will no longer want to apply independent judgment. Adherence to the rules of war will become less relevant as well.

With ongoing work in computers, artificial intelligence, robotics, and autonomous vehicles, we are attempting to make machines—inanimate objects—approximate the behavior of humans. At the same time, with work in soldier enhancements, be they physical, neural, pharmaceutical, or performance-based, we are attempting to make humans behave more like machines. In both cases, we are blurring the concept of what it means to be human. Soldier enhancements challenge our roles as moral agents, which require individuals to be held justifiably and absolutely responsible for the consequences of their actions. Enhancement technologies intended to modify the soldier to kill more efficiently and survive in battle may diminish his sense of humanity.

Is the concept of pride, so important to unit cohesion and esprit de corps, still applicable? Soldiers who train together, who fight and die and survive together, develop strong bonds.

Soldiers who used to train individually and with units to succeed in combat may now get their capabilities and courage artificially through pharmaceuticals or other medical interventions. Can soldiers under the influence of behavior-modifying drugs or electronics be held to account for their actions? If the soldier is using performance-enhancing drugs, we can question what sense of accomplishment he can feel; if drugs enhance his cognition or reduce his fear, how does free will enter in? The philosopher Arnold Toynbee has written that "it is characteristic of our human nature that we rebel against our human limitations and try to transcend them." In 1964, he couldn't have imagined the types of artificial enhancements being considered today. Might a soldier who fears nothing unnecessarily place himself, his unit, or innocent bystanders at risk? A soldier's ability to consider the suffering of others will likely be affected. What about the impact of memory-altering drugs on the soldier's sense of guilt, which might be important in decisions about unnecessary and superfluous suffering? These are important emotions in war and they form the basis for many of the tenets of Just War Theory.

To review, the soldier of the future will be tied into a computer network where he can access vast amounts of data and where his physical location, health and performance status, weapons capability, and possibly state of mind will be continuously fed into large computer models and made available to senior commanders. The soldier will be a node in a large network, where decisions might be distributed, made at a higher level, or made by a machine. In any case, there will be a loss of individual accountability. If a soldier feels no accountability to his fellows, or his fellows to him, feelings such as honor and loyalty to the unit are likely to be absent. Intense reliance on computers results in loss of creativity and decision-making ability. We become too dependent on and addicted to machine aids and are unable to function as well when they are unavailable. With widespread modeling and simulation of combat engagements, we will have theoretical calculations of what should be, absent the reality and unpredictability of human behavior and environmental factors on the battlefield. At the same time, we will be removing the human context from decision making, possibly with lethal consequences.

My overall concern with the new technologies is that we are, seemingly with reckless abandon, rushing to incorporate computers, robotics, artificial intelligence, and other automation into ever more human activities, both civil and military. Manufacturing long ago gave way to assembly-line automation. We get great, inexpensive, quality products, along with growing worker displacement. In health care, computers and algorithms make excellent diagnoses, and robots perform exquisite surgeries, displacing highly skilled doctors. Police departments now depend heavily on data mining and predictive analytics to solve and prevent crimes, eliminating the need for human deductive reasoning. Self-driving cars are on the horizon. In war, that most human of endeavors, we are designing machines to handle every phase of conflict, including finding an enemy, tracking them, identifying their capabilities, targeting them with an appropriate weapon, and destroying them.

The issue is the speed at which this is taking place, and

how little debate or understanding will go into it. Mine are not the rantings of a Luddite. Most, if not all, of these developments have many positives. We gain much of value in the way of quality, safety, and security. But we are losing something as well. In the civilian world, work is fundamental not only to personal well-being but to self-worth. As the world's population continues to soar, we are rapidly making human work obsolete. Basic human capabilities such as mathematics, navigation, and rational deliberation and decision making are being lost, as are complex activities like driving or flying an airplane. As we cede these abilities to our machines, they atrophy in us. These are serious consequences for civil society. As for the marginalization of human capabilities in warfighting, it is simply an awful idea.

If current trends continue, war will become more a test of technologies than a struggle between humans. Either way, humans will die as a result. In the latter case, a human is deciding to take a human life, with all that entails. "A robot's targets do not have the option of an appeal to humanity the way they might if a person was behind the weapon," as the UN Special Rapporteur on extrajudicial, summary, or arbitrary executions correctly notes. Death by algorithm would be the ultimate indignity for a soldier.

We must be careful not to allow our love affair with technology to cloud our judgment or blind our thinking to the realities of war, or to lull military planners into the mistaken belief that the United States will always enjoy technological superiority. Echoing the concerns of senior combat leaders that the overdependence on technology will alter the ways soldiers think about soldiering, Lieutenant General H. R. McMaster has said that "the warrior ethos is at risk because some continue to advocate simple, mainly technologically based solutions to the problem of future war, while ignoring war's very nature as a human and political activity that is fundamentally a contest of wills."

Technology changes the institution of the military itself. Armored vehicles displaced cavalry soldiers and all of the logistics that accompanied them, only to replace them with a different organization, doctrine, and logistics. Intercontinental ballistic missiles created an entirely new force structure and set of rules of behavior. The aircraft carrier spelled the end of the battleship navy. The introduction of nuclear weapons changed the organization of the military and how it trained and equipped itself, resulting in entirely new doctrine and both strategic and tactical forces equipped to engage in nuclear combat. Militaries were subsequently designed to fight smaller, regional wars. Now we can see the military reorganizing to accommodate the growing dependence on cyber warfare, or creating entirely new units and doctrine for drone pilots. Along with these organizational and institutional changes will come changes in training and education and, indeed, changes in how soldiers think about potential enemies and their comrades-in-arms.

Too often, the dogged pursuit and implementation of technology solutions are not accompanied by comparable investigation of the potential consequences. This is true in civil life and especially in the military, where the implications can be fatal. Yale University ethicist Wendell Wallach has noted in his book *A Dangerous Master* that we may well be at an inflection point in our development of technology and our ability to cope with its implications. That is, we are at a crossroads where we could either lose control to the technologies themselves, or find a way to control them. I prefer to think of the issue as a problem of divergence. While technologies are becoming ever more complex, our ability and our desire to understand them is declining. Will we allow the divergence to continue unabated, or will we attempt to slow it down and take stock of what we as a society are doing? Responsible research and development organizations can enforce better planning and stronger reviews. Decision makers can demand more debate before committing to weapon systems.

With the rapid introduction of new technologies, attention to the laws of armed conflict is more important than ever. Employment of some of the more advanced technologies, and even the research, needs to be accompanied by careful thought about their ethical and moral implications. DARPA has recently begun to address some of the ethical, legal, and societal issues associated with its work. While the agency has enjoyed a long history of technology successes, it has also experienced some serious failures, like the ill-fated Total Information Awareness (TIA) program in 2003, which raised major concerns about privacy and civil liberties.

In 2010, DARPA sponsored a groundbreaking study by the National Academy of Sciences that delved into the ethical and societal implications of increasingly widely available technologies that could be used for both civil and military applications. I was a member of the study committee. Some potential questions or considerations that we identified for DARPA about advanced research were: Does the research or application compromise something essential to what makes us human? How will more scientific knowledge or better technology affect judgments about issues such as safety, fitness for human use, or precision of application? What is the nature of harm involved, if any, with a new military application? How will adversaries respond if they are the targets of a new application? How will we respond if we are the targets of a new application whose development we sponsored? What is the impact of an application on civil liberties, on economic relationships, on social relationships?

Among other relevant issues, the academy report discussed important topics such as the precautionary principle in research, wherein scientists must demonstrate in advance that their research will not cause harm; cost/benefit analysis, in which decision makers employ a consequentialist approach; and risk communication, in which responsible agencies provide adequate and useful information to the public or other important potential recipients of a technology to ensure their support. The study committee also made recommendations to DARPA about ways government research organizations could, with minimum disruption, regularly conduct analyses of the ethical and societal implications of their work.

Another member of the committee, Carnegie Mellon University social scientist Baruch Fischhoff, writing in 2014 about the report, commented on the notable absence of detailed recommendations for individual technologies, saying that "specific research situations require situation-specific analyses." He went on to say about the ethical principles that might be relevant to a research project that "sorting them out requires thoughtful analysis—by ethicists, spiritual leaders, and others, in conjunction with those affected by the project." DARPA does in fact call upon independent groups of ethicists to review their work in neuroscience, biology, and privacy and civil liberties. This needs to become the norm throughout government and industry.

While these questions may seem a bit distant from the concerns of a warfighter, they aren't. In the end, the military will figure out how to fight in the future in accordance with the laws of war. That is their business and they do it well. Still, the questions remain. While our forces will follow the laws of war, are they ethically equipped to judge the implications of their acts? Critically, will decision makers, and the public the military is sworn to serve, understand or care? Unfortunately, the prognosis is not favorable.

4

SOCIETY AND THE MILITARY

It ain't what happens here that's important. It's what's happening back there. Lieutenant, you'd hardly know there's a war on. It's in the papers, and the college kids run around screaming about it, but that's it. Airplane drivers still drive their airplanes. Businessmen still run their businesses. College kids still go to college. It's like nothing really happened, except to other people. It isn't touching anybody but us.

—Staff Sergeant Gilliland,
Fields of Fire, by James Webb

People don't know—and don't want to know—what you've been through. . . . There are no bond drives. There are no tax hikes. There are no food drives or rubber drives. . . . It's hard not to think of my war as a bizarre camping trip that no one else went on.

—Iraq War veteran

THE UNITED STATES has breathtaking technological capabilities and an amazing new arsenal of high-tech weapons. These new capabilities will bring tremendous military advantages, but are incredibly complex, differ from current weapons in fundamental ways, and will present ethical challenges to the soldier. The military professionals who must adapt to new weapons and new ways of war want to do so, while senior military planners and commanders publicly express concerns and their intention to consider ethical issues. Besides the military itself, though, who cares? Who understands the technologies and who understands or is concerned about the challenges our soldiers face?

The U.S. military has changed dramatically in the last century. Following the world wars and the disastrous Vietnam conflict, it underwent a serious transformation, with an ever-greater reluctance to engage in conflicts without clear objectives, a growing aversion to casualties, and an insatiable appetite for high-tech weaponry. The post-Vietnam period saw the military redesign itself, reorganize, and rearm with modern equipment and tactics. We became a technology powerhouse.

War and technology have defined our history. Americans spend vast sums on weapons and aggressively export our weapons to others. We assume our technological superiority, an assumption that is increasingly challenged and that has potentially dangerous consequences if we think it will allow us to impose our will on others.

Since the end of the Cold War, not only has the military transformed, but the separation between it and society has widened. The Vietnam War, fought as it was by a conscripted military, caused massive societal dislocations. When at the end of the Cold War two superpowers were no longer threatening to annihilate each other, we gloated about the triumph of democracy over communism, rather than building on the potential for peace. We saw access to the former Soviet states as an opportunity to encourage capitalism, not democracy and human rights. We involved ourselves in many small conflicts around the world. Our priorities were misplaced. And the public lost interest.

Today there is a woeful distance—a chasm—between the military and the society it is designed to serve. Questions of how to behave in war and the proper weapons to use, important to the soldier's view of himself as a professional and critical to how the nation wishes to be seen, seem to be of little interest to most Americans. Make no mistake: the willful ignorance of the American public and its leaders will have dangerous consequences. Most Americans, including many of our political leaders, pay scant attention to military issues until a situation arises concerning our armed forces. Then they act based on emotion and political expedience rather than on facts, and that rarely ends well.

ARROGANCE, EXCEPTIONALISM, AND HUBRIS

For decades, Americans have viewed technology as a panacea. Assembly lines, electrification, automation, computers, instant communications, better living through chemistry, and advanced diagnostics in health care all represent our deeply held dependence on technology to solve our problems. It is hard to argue with these advances. Wendell Wallach noted that "technological growth is a result of what we humans want from technology and our willingness to buy into the belief that technology can provide what we want and need."

British author Christopher Coker tells us that one of the lessons of Thucydides' *History of the Peloponnesian War* is that, unlike the Athenians, we should never allow ourselves to be seduced by our own power. Such was the case in the 2003 invasion of Iraq. Decision makers at the time assumed that with our overwhelming military superiority, the war would be over in days and the United States would be hailed and welcomed as a liberator. While the initial phases of combat were short, our arrogance, possibly born of our technological muscle, blinded us to the realities of the post-invasion difficulties. As a military and as a nation we must never take on a mantle of invincibility or imperiousness. Such a feeling of superiority often leads to military miscalculations with disastrous consequences, like the long nightmare in Iraq, or the disastrous foray into Somalia, or even President Reagan's insertion of Marines into Lebanon. Some senior military personnel urged better planning and understood the instability our invasion of Iraq would cause. They were ignored, and we are still living, and will continue to live for some time, with the awful fallout of that war. Along with unbridled military adventurism comes a loss of careful deliberation about the consequences of our actions.

Even with experience as our teacher, many continue to think that our technical military prowess makes us invincible. Far too many technology zealots believe that all problems can be solved by science, and military zealots believe that our amazing new weapons will allow us to win any war with few casualties. Many of the more strident supporters of the frequent use of military force insist that technology will give the United States an unquestioned superiority over all potential enemies. Such hubris, feelings of exceptionalism, and militarism all too often replace good judgment. Our awesome technology may even encourage us to engage in unnecessary and gratuitous violence. At a 1993 meeting with Joint Chiefs chairman General Colin Powell, then secretary of state Madeleine Albright asked, "What is the point of having this superb military that you're always talking about if we can't use it?" Many leaders and decision makers, emboldened by our technological prowess, frequently argue for more military resources and for far-flung demonstrations of American military might.

When the United States handily defeated Saddam Hussein in the 1991 Gulf War, the public could watch it live on CNN as if it were entertainment. There was a view among military planners that we were experiencing a revolution in military affairs, that American military technological advantages would eliminate the fog and friction of war. The following decade became the era of military transformation and computer-network-centric warfare. It was the military version of the dot-com era, and it similarly seemed to lack any serious thought, reflection, or sober judgment about the capabilities provided by the technologies and the wisdom of such a one-sided, technology-centric approach. I fear we are taking a similar, overly optimistic view of the hoped-for benefits of artificial intelligence and autonomous systems. I believe this hubris has also become part of the fabric of our society and leads to an outsized perspective of both the power and the appropriate uses of the military. When we have an unrealistic view of technology, we get uninformed decision making, with politicians too quick to pull the trigger.

The Gulf War gave us an opportunity to demonstrate all of the amazing technologies in which we had been investing during the previous decade. The 1990s saw a continuation of our fascination with technology, with military and civilian defense leaders touting U.S. dominance. After the attacks of September 11, 2001, and the subsequent invasions and wars in Iraq and Afghanistan, we introduced armed drones and massive new surveillance capabilities. For more than a decade the United States invested heavily in more exquisite sensors, more lethal weapons, and more intrusive intelligence methods. "There was a vision," writes journalist James Mann, "of a United States whose military power was so awesome that it no longer needed to make compromises or accommodations (unless it chose to do so) with any other nations or groups of countries." If we are invincible, the thinking goes, why should we even have to think about the ramifications of a particular action? Who indeed could call us to account? In writing about how the United States should respond to the attacks of 9/11, the conservative columnist Charles Krauthammer opined that "power is its own reward. Victory changes everything, psychology above all. The psychology in [the Middle East] is now one of fear and deep respect for American power. Now is the time to use it." The arrogance was palpable.

The American public and our decision makers continue to think technology will allow us to impose our will on others with little or no cost to ourselves. DARPA's Dr. Arati Prabhakar agreed that many senior decision makers lack any real understanding of those technologies, but often make decisions based on their promised capabilities.

In his book *Virtual War: Kosovo and Beyond,* Michael Ignatieff warns modern warriors against the "moral danger" they face if they allow themselves to become too detached from the reality of war. He writes that "we see war as a surgical scalpel and not a bloodstained sword and in doing so we mis-describe ourselves as we mis-describe the instruments of death." He concludes that we need to stay away from such fables of self-righteous invulnerability.

Little thought is given to the destabilizing nature of some of the technologies we develop and the predictably negative reactions of others. Former Russian military officials have told me that programs like the U.S. Air Force's Prompt Global Strike initiative, generally thought of as a nonnuclear strategic weapon, create fear and uncertainty among other nations. As our goal is military advantage, this is probably exactly what is intended by U.S. planners. However, such actions cause others to react with their own similar, or perhaps even more aggressive, programs. For example, Russia recently announced the development of a long-range autonomous submarine capable of delivering a nuclear weapon into U.S. harbors. Our actions often appear arrogant and create, according to many observers around the world, unnecessary strategic instability. Counter-productive and expensive arms races result. There is little evidence that such considerations are debated or even considered by senior decision makers.

British academic Sir Alistair Horne has written that for the ancient Greeks, hubris was the folly of a leader who through excessive self-confidence challenged the gods. It was always followed by a reversal of fortune and, ultimately, divine retribution. In talking about the dangers of arrogance and hubris, Horne describes the tendency of generals and nationalistic political leaders who experience military triumph to overreach, and for the next generation to inherit their arrogance and complacency, with disastrous results. Horne warns that "what we and our leaders need to understand is that the exuberance that follows victory all too easily leads to the wrong decision." From Athens to Afghanistan, leaders have learned—and forgotten—that lesson far too many times. History reminds us that hubris rarely ends well.

In the first episode of the HBO series *Newsroom,* the fictional cable news anchor is asked by a starry-eyed college student to describe why the United States is the greatest country in the world. To a shocked and stunned live audience, he replies that the United States isn't the greatest country in the world, but that it used to be. He then proceeds to list all of the metrics in which America is lagging other countries and the not-so-positive differences in the way we now approach our problems. Whether you agree with this fictional character's conclusions, his data, or even with the intentions of the show's writers is irrelevant. The point is that it is irresponsible to make trite assertions about American "exceptionalism" that are not supported by fact. We may all wish to believe our country is better than others, but just asserting that it is so is arrogant in the extreme.

There is a great debate in this country about what constitutes exceptionalism. On the one hand, there are those who equate it with raw power. We are without a doubt an extraordinary military power. Power is obviously important, because without it we cannot defend ourselves or others, and we cannot speak from a position of strength. Too often, however, we act like bullies. Our many "interventions" are evidence of that. Others, like myself, view exceptionalism as something far different, tied to our values, preferring to think of it more in terms of "soft" power, our ability to influence with good deeds, thoughtful behavior, and setting a good example, all backed up by our willingness to use the raw power we have, but only if we have to.

THE MEDIA

Where do the public and some of our leaders get their understanding of the military? Since the public is largely disconnected from the armed forces, there is little incentive for it to learn about what the military really is and does. Its knowledge is superficial at best, generally gleaned from news, television, movies, and the Internet. It is a skewed picture.

The advent of the twenty-four-hour cable news cycle in the 1980s radically changed the role of the media in news reporting. Driven heavily by competition and the demand for profitability, cable news outlets were forced to fill the time with content. While the viewer benefits from a constant awareness of breaking events around the world, such real-time reporting often lacks context. It is truly heartbreaking to see innocents being killed or injured in the Middle East or Ukraine, but such reporting is often devoid of background. Sad though it may be, the viewer is only being told of the event, not the causes.

If cable news was an important development, the Internet was an extraordinary one. At least with cable there was some possibility of editorial control. With the Internet, there is none. Organizations and individuals are able to post anything—true or untrue—for all to see. Cable news made it possible to tailor news for different interest groups. The Internet makes it possible for individuals to converge on an idea and amplify one another's beliefs, drowning out any contrary views.

We have now entered what some refer to as the "post-truth" era, characterized by the repeated assertion of talking points to which factual rebuttals are ignored. In the post-truth era, facts are irrelevant. As a strategy, it works because it allows people to dispense with critical thinking as their beliefs are constantly reinforced by others with similar views. Russian media official Dmitry Kiselyov noted that "the age of neutral journalism has passed." This is bad enough in the political arena. When applied to geopolitics and military affairs, it could have disastrous consequences, as falsehoods could lead to unnecessary acts of violence.

Much of what the public knows or thinks about the military derives from entertainment. Every year, movies romanticizing the military and conflict are produced and released to captivated audiences. They generally represent only one side of a conflict and glorify war. They depict unerringly accurate weapons (as do video games) and the morality and righteousness of our cause. Often, in an attempt to dramatize, moviemakers get things wrong. More than once I have had to explain to a senior decision maker that the limitations of orbital mechanics make it impossible to "fly" a satellite to an area of interest to take a picture. Serious scholarly books are written about war and its consequences, but the vast majority of the public ignores them.

For decades, the media depicted war as heroic and failed to show its awfulness. *Apocalypse Now, Platoon,* and other films born of the Vietnam War depicted the insanity and brutality of the war, but focused mostly on the unseemly politics of that conflict. In the war on terror, highly trained and supremely competent special operations forces carry out raids in far-flung areas of the world. Americans are rightfully proud, but again unaffected. Movies are made about secretive operations and heroic American snipers, and the public gets to vicariously experience "war" in an air-conditioned theater. Today, stories about special forces and heroic Delta Force or SEAL teams are all the rage. A society that has no direct knowledge of the military and only experiences it through the media will develop a very lopsided, romanticized, and aggrandized view of it, and of war in general. Since people feel that their quotidian lives aren't really affected by it, they have no incentive to probe deeper.

The media and the public, with strong encouragement from service members, have in recent years been fond of discussing warriors in terms of their ancient progenitors, the Spartans. As former military intelligence officer Jim Gourley describes in *Foreign Policy,* modern military culture has crafted a whole mythology of Spartan life to validate highly romanticized beliefs about who the Spartans were. The danger, he says, is that American military culture unquestioningly accepts that mythology as fact, and in pursuing the idea of "the American Spartan" is becoming something that is neither Spartan nor American. Gourley points out the irony that Sparta was utterly bereft of architecture, literature, art, or science and that though much of our military inheritance comes from it, our politics and culture are all derived from Athens.

One disturbing trend I have noted is the tendency of decision makers in the Defense Department to filter their choices of "news" based upon what they want to hear. Mirroring a trend in society, military leaders, at least in my experience, favor one cable news channel over all others. In numerous visits over the years to so-called E-Ring offices in the Pentagon, where senior military and defense civilians reside, I have found that the television was always playing Fox News. If there was an additional screen, CNN was showing. Given the well-documented conservative leanings of military personnel, the television preference demonstrates once again that the Pentagon often functions as a big echo chamber. Balance is needed, especially in the military.

For those who have never had to fight, it is easy to call up notions of heroic soldiers taking an objective while the enemy cowers in fear. That is a dangerously fictionalized version of reality. There is a propensity in the American public and the media to refer to all American troops as heroes. In reality, many active-duty soldiers and veterans bridle at the term. Some who have seen significant combat told me they find its use distasteful, feeling it demeans those who have in fact demonstrated heroism under fire. One of my students, a former Marine who had not seen combat, complained that it made him feel like a fraud. As Marine pilot Carl Forsling noted, "Just being exposed to danger or enduring some degree of sacrifice doesn't make one a hero." Referring to frequent public recognition of veterans as heroes, he said, "Many of the people standing for applause put their lives on the line about as much as a 7-Eleven clerk in a marginal neighborhood." In the words of retired Air Force lieutenant colonel William Astore, who challenges the increasingly popular labeling of service members as heroes, "A snappy uniform—or even dented body armor—is not a magical shortcut to hero status."

DELIBERATE IGNORANCE

As we have seen, the public will no longer have the luxury of ignoring or paying only cursory attention to conflicts in distant lands. The question will be whether our decisions about resorting to war will be thoughtful, informed, and rational or based on knee-jerk, uninformed opinion. Recent terrorist attacks in America and Europe have shown that terror knows no borders. These are events that will affect us all, and we will need to rise to the occasion, making the right decisions based on facts and not the myths we tell ourselves. Not doing so may endanger us physically and lead to catastrophe. We cannot allow our technological dominance to lead us into messy and unwinnable conflicts overseas. Our real or assumed technology superiority cannot be the overriding factor in a decision to go to war. We should enter into conflicts only after serious deliberation, given the enormous human and financial treasure we will put at risk.

Do we even have the capability, much less the will, to deliberate seriously on issues of national importance? Advocating for military force is often the default conclusion for Americans informed by the media, and sadly, it is often also a preferred choice for decision makers who have little patience for or capability to understand other, more complex options.

Surveys have consistently shown the desperately poor state of scientific literacy in the United States. We rank lower in science, technology, engineering, and mathematics than many other countries. The United States now ranks twelfth in the number of twenty-five- to thirty-four-year-olds with university degrees and fifty-second among 139 nations in the quality of our mathematics and science instruction. Our health care system, the most expensive in the world, consistently reports mediocre outcomes. With health care costs consuming almost 18 percent of its GDP, the United States is last, or near last, in measures of access, efficiency, and quality compared to eleven other industrialized nations. The study of foreign languages and foreign cultures, once required in U.S. schools, is now just a quaint idea. In 2009, only eleven states required any language study at all as part of K–12 education. Our education system, while including some of the best universities in the world, fails a reprehensibly large number of our citizens. And if you ask a random citizen, they wouldn't know an Air Force squadron from an Army company.

Sadly, there are those who not only don't know, but refuse to know. The science deniers, including not only millions of citizens but far too many politicians—"leaders"—deny climate change, evolution, and the efficacy and safety of vaccines. Regardless of overwhelming, irrefutable scientific evidence in support of these theories, some people refuse to accept that human activity is harming the environment, a sitting congressman can refer to the theory of evolution as "lies from the pit of hell," and far too many parents deny their children protection from potentially fatal diseases, choosing to believe instead that vaccines cause autism, despite a complete lack of corroborating evidence.

In regard to foreign affairs and military issues, only an embarrassingly small percentage of the public knows the difference between Shia and Sunni Muslims, and precious few can even locate Syria, Libya, or the Spratly Islands on a map. What they know, and wholeheartedly believe, about these issues is what their favorite cable channel commentator tells them to believe. Most citizens have no sense of the brutality and hellishness of war and of the ethical challenges faced by soldiers in harm's way, and simply seem to think that we have the military might to impose our will whenever we wish. Many feel that anytime the United States is challenged in the world we should "do whatever it takes" and should practice "an eye for an eye." Action is the default condition. Deliberation takes too much time.

There is a strong strain and long history of anti-intellectualism in American culture. In its current form, it dismisses science, the arts, and the humanities in favor of entertainment and self-satisfied ignorance. Throughout history, anti-intellectualism has also frequently led to violence. In Thucydides' *History of the Peloponnesian War,* Diodotus argued against vengeance and haste in putting to death an entire city, saying that those who make wise decisions are far more formidable to enemies than those who rush madly into strong action. The anti-intellectual Cleon argued for swift action and perfunctory discussion, deriding deliberation, saying that men of action are better leaders. International security scholar J. Peter Scoblic notes that one of the most important presidential leadership qualities is "knowing how and when to do nothing," or "knowing when to show patience, to tolerate delay and to restrain the urge to act."

William Manchester, in his book *A World Lit Only by Fire,* discussed the fate of educated men in the time of the Protestant Reformation, during which the joys of illiteracy were extolled. Intolerance, contempt for learning, the burning of books, even death at the stake awaited humanists and learned men. In the current Internet age, anti-intellectuals create a culture in which, as Canadian author Ray Williams describes it, "every fact is suspect, every shadow holds a secret conspiracy. Rational thought is suspect. Critical thinking is the devil's tool." Williams refers to these anti-intellectuals as the "metaphorical equivalent of an angry lynch mob." In America, anti-intellectual attacks were leveled against no less a figure than Thomas Jefferson. His critics believed that intellect made men timid and ineffectual, that intellectuals were likely to vacillate rather than act, and that they often favored abstract, radical, or even "foreign" ideas over the quintessential American values.

One of those quintessential American values has, at least recently, become the lionization of the military. We put ribbons on our cars, we give soldiers discounts at shopping malls, we stage elaborate performances at sporting events, and country-and-western singers even create songs extolling the virtues of our warriors. The American people thank soldiers for their service while thanking their lucky stars that they don't themselves have to serve. They are reticent to criticize the military, lest they be viewed as unpatriotic and ungrateful. The military has for too long been a thing apart in our nation. We pay lip service to it. There is no requirement to participate in it, or pay for it. There are no sacrifices. In the 1980s, President Reagan promised Americans that they could have a robust national defense *and* an expansive lifestyle, with limitless economic expansion. History has shown the folly of that concept.

We remove ourselves from the idea of what war really is and delude ourselves with clichés and Hollywood images at our peril. Combine this with politicians waving the flag of American power with simplistic assurances that we can "set the sands of Iraq on fire" or "blast ISIS to kingdom come" or apply "shock and awe," and we have a nation at alarming risk of self-destruction born of ignorance and delusion.

A PUBLIC-MILITARY CHASM

I believe that the individual citizen, largely unaffected by war, is intellectually too complacent and too self-absorbed to care about issues of technology and war or their consequences. Thinking is hard work and takes too much time. I am stunned and depressed by the almost complete lack of knowledge about military matters and military people by the vast majority of those with whom I come in contact. Once, as I stood in uniform in the lobby of my hotel, awaiting the arrival of a colleague, someone approached me and asked if I would take his luggage to his room. In a major metropolitan airport, a couple approached me to thank me for my service and asked me if I was a sergeant. I was a brigadier general. These are harmless incidents of ignorance. More troublesome are the frequent questions and comments which imply that service members are either unthinking automatons, or even that a job in the military was the only one they could get.

Many senior decision makers see the military as a tool, virtually ignoring the impact of frequent deployments on soldiers and their families, and doing little to support their reintegration into society upon their return from the war zone. Since the beginning of the all-volunteer military, senior leaders have increasingly resorted to military deployments and the use of force. They receive little blowback, economically or politically, when they do. In the seventy years since the end of World War II, U.S. forces have been deployed sixty-five times. For better or worse, our military leadership calls the shots, improperly guided by congressional oversight itself driven overwhelmingly by partisan politics. Only one-fifth of the members of Congress have served in the military. The number of elected officials who are veterans has plummeted to its lowest point since World War II. Members of Congress are quick to say they support the troops and veterans, but allow petty disagreements and ideology to get in the way of constructive action. Meanwhile, our troops suffer. Administration and congressional inaction on the shamefully substandard service our soldiers receive at Veterans Administration hospitals is a national disgrace.

Soldiers are increasingly critical of their leaders' propensity to resort to war. The Vietnam War created skepticism and cynicism that are with us to this day. Just one-third of veterans say that the wars in Afghanistan and Iraq were worth fighting, and another one-third say neither has been worth the costs. Iraq combat veteran Patrick J. Murphy said, "If more leaders knew firsthand the costs of war, they wouldn't be so quick to inflict that experience on the generation of young soldiers that follows them." Even the soldiers fighting and dying often don't know why. "What was the point?" one asked. "I risked my life and I got shot, and I got blown up—for what?"

Military concerns are just not on the minds of most citizens. Eighty-four percent of post-9/11 veterans say the public does not understand the problems faced by those in the military or their families. In 2011, the Pew Research Center surveyed Americans about their connections to the military and found a considerable gap: "Never has the U.S. public been so separate, so removed, so isolated from the people it pays to protect it." A growing chasm between the U.S. military and the public it serves causes a skewed understanding of the role of the soldier on both sides.

In the 1950s, C. P. Snow wrote eloquently about what he called *The Two Cultures.* His subject was the distance between the arts and humanities and the sciences. Today in the United States we have a different and far more dangerous "two cultures" problem. We have the military and everyone else. Without these two cultures mingling, without communication and integration, there will be dire results.

Citizens need to understand the importance of when and how and according to what rules we fight. What are the effects of all of the changes in technology and the ways of war on the society from which the military is drawn? What do these changes mean to the average American? Are our decision makers sufficiently sensitized to the implications of the new ways of fighting and how these new technologies affect their own decisions about going to war? How are the attitudes of individuals and civilian institutions shaped by the wars we fight and the means we use to fight them? Conversely, how do the attitudes of civilian society affect how the military behaves? The answers to these questions are critical.

For the United States, which prides itself on liberalism (in a broad, not political sense), the military should be a reflection of those deeply held values. How and when our forces are sent to war or engage in worldwide conflicts and the means they use to achieve victory should demonstrate the values we hold dear. For it to be otherwise represents a serious detachment of a populace from its protectors. New technologies and ways of fighting appear to make war easier, but easier does not equate to capricious or less serious.

There are two prevailing arguments about the "all-volunteer" military. The first argument, with which I agree, is that it is dangerous to have an armed force that is so skewed demographically that it doesn't reflect the society it serves. Such a narrow demographic might also have similarly narrow views on any number of issues, including human rights and fairness. The opposing argument is that for the military to reflect society is unimportant. As long as the force is professional, efficient, and overwhelming, this thinking goes, it will do and behave as directed by higher command authority. The presumption is of course that civilian leadership of the military has the necessary sensitivities to direct and control ethical behavior from its soldiers. I believe this is faulty thinking.

As the military has evolved into a far more professional and highly trained force, the U.S. populace has increasingly focused instead on the economy, the Internet, social media, entertainment, and other personal and interest-group issues. While the military has been sent to fight in a series of small conflicts in Haiti, Somalia, Bosnia, and elsewhere, and big conflicts in Iraq and Afghanistan, the American people unrelated to the military have been unaffected. None of the conflicts in which we've been involved, especially those that occurred after 9/11, have required any sacrifices by other Americans—no conscription, no rationing, no new taxes. In fact, we have seen the rise of Generation X, Generation Y, and the Millennial generation, whose interests are very much unlike and even antithetical to those of typical soldiers. In a recent survey, 60 percent of

Millennials polled say they support committing U.S. combat troops to fight ISIS, but 62 percent say they wouldn't want to personally join the fight, even if the United States needed additional troops.

Increasingly, military members are drawn from a narrowing slice of American society. Minorities represent a larger percentage of the military than they do of the American public. Only about one-half of one percent of the U.S. population has been on active military duty at any given time during the past decade and a half of sustained warfare. Retired Army general Karl Eikenberry and David Kennedy of Stanford University describe the situation as a "disturbingly novel spectacle: a maximally powerful force with a minimum of citizen engagement and comprehension." It is important, especially as the types of conflict change and new, advanced weapons technologies emerge, that potential future leaders and the broader American public be exposed to military and technology issues and what they mean for the future. Unfortunately, the American military is like the 911 system: we call only when we're in trouble.

Retired admiral Mike Mullen, former chairman of the Joint Chiefs of Staff, expressed concern about the growing disconnect, saying, "fewer and fewer people know anyone in the military. It's become just too easy to go to war." If more Americans were affected by military matters, politicians would have a harder time starting unnecessary wars and continuing them long past when they should have ended.

A chasm, however, always has two sides. Former defense secretary Robert Gates expressed concern about a statement by a Marine sergeant that the military has better standards and values than civilians. Gates worried that the military would become too separate from and deem itself superior to the society, the country it is sworn to protect. In a 2003 poll of military members, approximately 60 percent of the respondents said they felt the military had moral standards superior to those of civilian society and classified the moral fabric of America as fair or poor. Whether such an assessment is accurate is a matter of opinion, but the sentiment itself is dangerous and troubling. Army chief of staff General Martin Dempsey has expressed his concern about the dangers of having a military with an inflated sense of superiority.

If the public doesn't know or understand or, worse yet, doesn't care about how and when its military should be employed or how it should behave, the use of military force will become too commonplace, too easy. If members of the military know the public doesn't care or understand why they do what they do, they might develop their own reasons, or be less apt to care about it themselves. A chasm of knowledge and understanding could easily become one of values and behavior. If the military becomes too insular, too self-righteous, and too independent, the public could quickly turn against it.

The military doesn't want war. The public needs to understand that. General Douglas MacArthur said that the soldier above all others prays for peace, for it is the soldier who must suffer and bear the deepest wounds and scars of war. Our senior leaders and the public also have responsibilities. The military should not have to bear the burden of the nation's wars alone.

5

WHERE DO WE GO FROM HERE?

We seemed about to enter an Olympian age in this country, brains and intellect harnessed to great force, the better to define a common good. . . . It seems long ago now, that excitement which swept through the country, or at least the intellectual reaches of it, that feeling that America was going to change, that the government had been handed down . . . to the best and brightest of a generation.

—DAVID HALBERSTAM, *The Best and the Brightest*

OUR NATION HAS A SERIOUS PROBLEM. Technology is moving so fast that few can understand it. What Americans think of as war is a thing of the past. The future will be at times frighteningly, almost unimaginably different. Our soldiers are overwhelmingly good and decent, but future war and weapons are going to place enormous burdens on them. Americans are largely unaffected by issues of technology as they relate to war. As long as that remains true, they are content to let politicians use the military as a tool and deploy it for reasons that mostly have little to do with our security. We spend vast sums of money and commit American soldiers' lives to unnecessary conflicts—wars fought for spurious reasons or for political expediency—with little public debate or concern. That cannot continue. Technology and war have always been closely coupled, but the radically different types of weapons and new types of conflicts will demand the active engagement of soldier, decision maker, and public alike.

We are at an inflection point, a time when important decisions must be made, and we may not get a second chance. The future of warfare is coming fast. It is going to be so complex, raising such complicated questions, that if jingoistic or ill-informed politicians speaking platitudes and clichés are the only decision makers, we may find ourselves in wars we should not wage and cannot win. If the American people want to sit at home waving the flag but ignoring the way we fight our wars and the people who fight them, they may find that such complacency comes at a terrible cost. The muted, almost insignificant response of most Americans to the Iraq War and its horrible cost in lives and money tells a heartbreaking story about the public's priorities and its true regard for the military. If such a stunning display of dishonesty by our leaders could not generate more outrage, could anything, ever?

We are a great nation and we have shown over and over again our ability to rise to great challenges. Our preparation for future war will demand the dedication of brilliant minds. Who is going to step up and lead the nation through it? The speed and complexity of new weapons and our diminishing ability to comprehend them, combined with the lack of public engagement with military and technology issues, paints a very dark picture for the future. It is a picture we have to redraw.

Everyone involved in this growing nightmare is operating with a deficit of knowledge. The public knows little about technology or war. Soldiers live in a special world and often find it hard to understand civilian society. Decision makers are inadequately knowledgeable of the military and are driven by politics. Weapons consume an enormous amount of our national wealth. The situation is serious and numerous writers have presented and debated ideas on what to do to resolve it. Some steps would be more difficult than others, but some require nothing more than political will.

We should begin by clarifying our national policy. Every president issues a national military strategy and sets the tone for the rest of the national security establishment. The strategy should not simply be a rationale for committing forces to combat; it should be a comprehensive plan for true national security. When our nation is actually threatened, then "national defense" might trigger a war decision. The idea of a preventive war, however, based on some ill-defined threat to "national security," should be unequivocally and explicitly abandoned.

For decades, Congress has deferred to the president in matters of war, shirking the responsibilities given to it in the Constitution. A national military strategy should also clarify the decision-making process and would have to result from a resolution of the long-simmering issue of war powers between Congress and the executive branch. "The president has been commander in chief since 1789, but this notion that they can go to war whenever they want, and [ignore] Congress, that's a post–World War II attitude," says Louis Fisher, scholar in residence at the Constitution Project, a nonprofit Washington, D.C., think tank. Congress and the president refuse to openly engage on thorny disagreements on war powers, preferring instead to kick the can down the road and keep their options open. We must hold our decision makers accountable. At a minimum, any involvement of U.S. forces in combat operations should be accompanied by an Authorization for Use of Military Force (AUMF), a congressional resolution of approval for the president to order troops into combat. It was first used after 9/11 when Congress authorized operations against al-Qaeda. One of its main values is that politicians must be forced to go on record with their votes. Conor Friedersdorf, writing in *The Atlantic,* says, "Taking a vote in favor of war, or against it, is a perilous act. They're declaring themselves on a subject of great consequence. If they're proven by later events to have judged poorly, they can be held accountable. As a result, many legislators abdicate their Constitutional responsibilities on matters of war and peace."

Politicians and decision makers routinely lie or tell half-truths to the American people about military issues and war, and they get away with it. We were dragged into the Vietnam conflict based on a questionable "attack" on a U.S. spy ship, resulting in the sweeping war authorization known as the Gulf of Tonkin Resolution. We were lied to about the reasons for going to war in Iraq, the United Nations and the American people were presented with questionable intelligence, and we invaded a country that presented no imminent threat to us. We must demand honesty in our leaders and punish them when we don't get it. Politicians cannot be allowed to lie to the public about reasons to go to war. They talk endlessly about "national security" and "protecting our freedoms" but never explain how our foreign adventures are making us safer. Military forces are given near-impossible tasks often with inadequate resources and no one is held responsible. We are so enamored with military capabilities that we have experienced repeated budget crises in recent years because many in Congress could not bear the idea that social programs might be funded at a level equal to or greater than weapons and the military. We accept mediocrity from our government officials, and that's what we get. We must demand better. The stakes are high indeed.

Politicians and decision makers are deluding themselves and the public about the true cost of war. Since September 11, 2001, the military has depended heavily on Overseas Contingency Operations (OCO) funding to pay for its wars in Afghanistan, Iraq, and elsewhere. Were these funds not available, it would be impossible for the military to monitor unanticipated expenses and budget properly for normal annual expenses, including the cost of equipment, maintenance, and research. The accounts are necessary, but they are regularly abused.

Since 9/11, these funds have been used for a variety of purposes having nothing or very little to do with actual overseas operations. Many government organizations, including ones in which I served, see them as an opportunity to purchase additional systems and equipment with only tenuous connection to the actual wars. In 2009, lawmakers tried to use OCO funds to acquire eight additional C-17 transports that the Defense Department didn't want and more F-22 fighters though they had never seen action in Iraq and Afghanistan. The funds are improperly considered "off-budget" and are not taken into account in congressional debates about the proper balance between military and domestic spending. That is a true outrage.

Congress was proud of itself when several decades ago it started requiring offsets to pay for anything new. In 2011 and 2013, it almost brought the economy to its knees over the question of a debt ceiling and whether or not we would actually pay the bills Congress had already incurred. Debt and a balanced budget have nothing to do with a sudden true military necessity. Sometimes, shockingly, taxes *do* have to be raised, and many presidents have done what needed to be done. If the American public is unwilling to cover the cost, the military would be restrained. We would quickly find out whether the people supported such operations.

There should be a broad national debate over the issues of the budget and how much defense spending is enough. The public must be a part of that debate and must understand it well enough to reasonably participate. The American public understandably wants to provide the troops with the very best equipment to fight our wars. It is amazing, though, how much of the defense budget is wasted, and how little the public knows about it.

The military annually presents its wish list for new weapons costing billions of dollars, claiming, truthfully for the most part, that the old weapons will put our forces at risk. Often, however, new weapons or upgrades are a way for the defense industry to continue to employ its workforce. The DOD accepts this, Congress accepts this, and the public is either ignorant of it or doesn't seem to care. Worse yet, a large number of systems are mismanaged and far exceed their budgets and schedules. The Government Accountability Office (GAO) reports that for seventy-eight major programs examined in 2014, procurement costs were 46 percent over budget.

Another GAO study reported that of ninety-five programs studied, 30 percent were more than two years behind schedule. The weapons often end up being canceled due to the delays, depriving warfighters of needed capabilities. Given the massive government weapons bureaucracy and the many competing interests, it is not surprising that we have such frequent and large cost overruns and schedule slippages. There is ample blame to go around, and Congress deserves a big share of it. The lack of public outrage is downright disappointing.

It is one thing to dedicate precious tax dollars to effective weapons, and the public is happily willing to do so. It is quite another to see those dollars wasted. The deliberations of the congressional Armed Services Authorization and Appropriations committees should be opened up to far more scrutiny than they currently receive. There are think tanks for virtually everything these days. The finest minds we have should be working on these issues every day and effectively conveying the information to all of us. If the public could see some of the inane reasons for weapons purchases, it might be more apt to voice its opinions. Former U.S. Air Force chief of staff General Ron Fogelman cited the blatant "horse-trading" at the Pentagon on weapon system requirements as one of the reasons he retired before the end of his term. He was disgusted by a process in which weapons budgets were allocated to the services based less on actual need and more on a sense of entitlement to a fair share.

If we continue to spend astronomical sums of money on weapons for which we get less and less in return, while ignoring or underfunding the basic human needs of our people, we may find we have a first-rate military and a second-rate society.

President Eisenhower, in his oft-quoted final address to the nation, warned not only of undue influence of the military-industrial complex, but of the imperative to care for human needs, saying that "each proposal must be weighed in light of a broader consideration: the need to maintain balance."

Our leaders also must be honest with the public about the size of the military. The use of private contractors has gotten out of hand. In the Iraq and Afghanistan wars, those contractors sometimes outnumbered the active-duty soldiers. Contractors are expensive; they are often paid far more for doing the same job as an enlisted soldier. The overuse of contractors is not just confined to the battlefield. In many defense organizations, private contractors outnumber government employees, doing work that is of course not categorized as inherently governmental. In an amazing series of articles in the *Washington Post,* Dana Priest and William Arkin estimated that since September 11, 2001, of the 850,000 top secret clearances, 265,000 belonged to contractors. They also estimated that in the CIA, contractors make up a third of the workforce, or about 10,000 people. In fiscal year 2015, the costs for contracted service personnel, those not working on classified contracts or weapons manufacturing, was $114.8 billion. While this number has dropped in recent years, even at reduced levels it remains enormous. The 561,000 service contractors, who do everything from lawn mowing to computer support and engineering, still outnumber the entire active-duty U.S. Army of 475,000 soldiers. This is an area that absolutely demands more scrutiny.

We have shown with international nuclear arms control agreements that it is possible to limit the growth of nuclear weapons. Indeed, we have seen dramatic reductions. The major powers thereby sent a powerful and encouraging message. New warfighting technologies are a thing apart from nuclear weapons, but the world's developed and developing nations could take important steps in controlling the export of these technologies and the use of weapons employing them. We should be looking at arms control, technology control, and nonproliferation concepts, much as we have today with the International Traffic in Arms Regulations (ITAR) and Munitions Control List (MCL) organizations. The United States should take the lead in proposing such treaty agreements with other advanced nations.

We should find ways, short of conscription, to increase public involvement in the military and in military matters. As I have said, the public is disengaged from the military and there is no incentive for it to behave any differently. When the public doesn't know or care, there is little incentive for its representatives to know or care.

Admiral Mike Mullen has said that the next time we go to war the American people should have to say yes. One way to accomplish that would be to further lower the number of soldiers on active duty and place more burden of foreign deployments on the National Guard and Reserves, to be deployed only in a national emergency. In the case of such an emergency, the American people would have to be inconvenienced. They'd be involved, and they might think twice about allowing our decision makers to get us into another war.

Secretary of Defense Ash Carter recently proposed sweeping new changes to the military personnel system, the likes of which we have not seen since the Goldwater-Nichols Act of 1986. These new changes will allow far greater flexibility in allowing military personnel to move between military and civilian life and will allow outside experts, like technical experts from industry, to enter into service at more senior levels. This might be perfect for cyber specialists or specialists in synthetic biology who spend years honing their expertise and must return to industry or academia to maintain it. Such a plan has the benefit of not only increasing the pool of capable personnel, but also of leavening the military mind-set with fresh civilian thinking. Unsurprisingly, many old-time military personnel who "came up through the ranks" are skeptical. Congress has tepidly embraced the idea. Not only should these new ideas be encouraged, but they should be radically expanded to other government organizations such as the Department of State. If the government is to get the best and the brightest, it must think and behave differently.

Over time, the involvement of the military in operations other than war has grown to include peacekeeping and nation building, unfamiliar territory for soldiers. In 2010, Secretary of Defense Robert Gates and Secretary of State Hillary Clinton jointly proposed a State Department reserve force, similar to the military reserves, that could be used in reconstruction and stabilization operations in foreign countries. The creation of such a force would have had a double benefit, removing some of the burdens on the active military and at the same time involving more civilians in important foreign operations. This seemed like a great way of spreading U.S. influence overseas without deploying military forces. While Congress failed to fund this activity adequately then, it is an idea that should be reconsidered.

We must find a way to better educate the American people and their leaders about issues of weapons and war, recognizing where our education is lacking and taking even modest measures to fix those deficits. To start, our national leaders must consistently, frequently, and vocally value education and intellect.

In the 1950s, Congress passed the National Defense Education Act, which created the National Defense Student Loan (NDSL) program. This program was clearly in response to the Sputnik crisis and was aimed at encouraging students to enroll in universities to study science and engineering. I was a beneficiary of such a loan, without which I could not have continued my education. As an extra incentive, those who served in the military after college could defer for several years before repaying the loan. Those who taught in secondary education were also annually given a small percentage of loan forgiveness. With the on-again, off-again shortages of technical talent, there have been attempts in the military services to provide incentives for students in science and engineering, but alas, the military just didn't appeal to very many college-bound kids.

While military academies train the professional soldier, the Reserve Officer Training Corps (ROTC) prepares a larger and much broader segment of the college population. This program could be modified and expanded, perhaps to every college campus, where a small subset of it could be made a required course for all students. Whether attendees end up entering the military or not, such a program would educate a broader swath of the population about military matters. It would not be unreasonable to require this of any university receiving federal funds, and it should not be controversial, because only those students receiving scholarships from and wanting to enter the military would do so.

Wherever possible, military people pursuing advanced degrees should attend civilian universities rather than their own service colleges. This would increase the exposure of the civilian students to the military ethos, issues, and concerns, as it would for administrators and faculty. Some have even argued that it may be time to conduct all officer education at civilian institutions, with perhaps a single year at a redesigned military academy to cap the education of an officer prior to entering active duty. The idea has significant merit. The continuation of any postgraduate education in military schools is a relic of the past. As a mid-grade officer, I satisfied my requirement for so-called senior service school by attending Harvard University's John F. Kennedy School of Government. Unlike the services' military colleges, Harvard is clearly not an echo chamber for the military. There, I was able to take courses in a much broader range of topics than at a military school and to interact with officers, graduate students, and faculty with widely divergent views of politics and government. This should be the norm, not the exception.

The government could take the concept of the National Defense Student Loan program a step further, creating a new program of guaranteed low-interest student loans and broadening the areas of study to include any of interest and use to other departments of the government, like the State Department, the Labor Department, or even Congress and the Supreme Court. Political science, international relations, labor relations, and perhaps even business might qualify. To get such a loan, a student would have to agree to work *somewhere* in government service, perhaps for as little as one year, with repayment deferments and partial loan forgiveness offered for good measure.

Expanding further on the idea of government service, there have recently been calls for a national service program. Such programs have been promoted by retired Army general Stanley McChrystal and former secretary of state Hillary Clinton. The Peace Corps and AmeriCorps are two successful programs in which citizens can serve U.S. interests overseas and at home. Because the programs are voluntary, however, participation is limited. Just like the all-volunteer military, the percentage of citizens involved is a tiny fraction of the eligible population, so the programs are not exactly of national scope in their execution.

Far more impactful would be what Marine veteran Eric Navarro calls "a minimum shared civic experience" in which each eligible citizen would contribute a year to an agreed government program. In turn, they would become eligible to receive many of the benefits currently afforded citizens as entitlements. Strictly speaking, such a program would not be called mandatory. An individual could choose not to serve and not to receive benefits. In addition, citizens would have a range of choices on how they could serve. *Time* magazine's managing editor, Richard Stengel, proposed a somewhat less stringent version of this idea, suggesting that individuals committing to a year of public service be awarded a substantial U.S. savings bond in addition to their normal stipend.

The overwhelming focus of the U.S. media is not on important topics like science, but on entertainment, celebrities, and sports. When the media does attempt to cover important technology developments, it tends to hype them and often gets the facts wrong. Media coverage of the military, especially on commercial television, devolves into either breathless admiration for our "soldier-heroes" or antiwar rants. Rare is the reasoned discussion of the origins of a conflict, rationale for U.S. involvement, or the future consequences.

The media is of course market-driven and not amenable to direct government action, but we must demand more balanced and informative reporting. Independent, publicly supported media, not beholden to advertisers, needs to be greatly expanded, with government support, and not come under repeated attack as it has in recent years. A vibrant and independent press has always been an important element of our democracy. Accurate, unbiased, and experienced journalism cannot be allowed to disappear and must be properly valued. War is serious, and entertainment is not what our founders had in mind when talking about the need for a healthy free press.

As I've said, there is a host of legitimate reasons to pursue a broad menu of technologies for our armed forces, among them a desire to reduce both combatant and noncombatant casualties and to gain a qualitative technological advantage. However, the research and employment of some of the more advanced technologies, and especially the underlying rationale for their use, must be accompanied by careful thought about their ethical and moral implications, and the national consequences of a decision to resort to force. It is possible, with a reasonable amount of effort, to develop military technologies in a deliberate, ethical way and to begin to understand the consequences of using them, if we decide and commit to doing so. Critical to the success of such efforts, however, is that our leaders, and the public, must be willing to educate themselves, take ownership of the issues, and make deliberations on them a priority.

The situation in which we find ourselves did not develop overnight, and changing such deeply ingrained societal behavior will not be easy. Lasting change could take a generation, but there is too much at stake to delay. Inaction is not an option.

A CONCLUDING PLEA

As our case is new, so must we think anew. We must disenthrall ourselves, and then we shall save our country.

—Abraham Lincoln

RIGHT OR WRONG, American soldiers have been continuously engaged in some form of conflict for decades, and that is unlikely to change. The modern milieu is a toxic brew of global instability, economic upheaval, political polarization, and rapid technological change on a scale not seen in several generations, perhaps ever. Previous generations did not have to deal with rapid globalization, instant communications, social media, artificial intelligence, and the explosive rise of developing nations. The challenges are unprecedented.

The horrific attack on September 11, 2001, impelled a national desire for vengeance and unfortunately set off a cataclysmic series of events, including the ill-considered and disastrous invasion of Iraq. After the attacks, the president declared a national state of emergency. I signed orders recalling reservists to active duty and extending active duty for those whose tours were due to end. I know of soldiers who lost promising civilian careers after repeated deployments, and have little doubt that some of those I ordered to serve were injured or killed in combat. In stark contrast to the sacrifices of our military forces, the American public was given an extra tax rebate and told to go on about its business as usual. The all-volunteer military, including members of the National Guard and Reserves, was sent over and over again to fight. There were no new taxes, no draft, no impact on the average American. There was something horribly inconsistent about the president declaring a state of emergency when only the small segment of our population in uniform was affected. No wonder it was easy for Americans to "support the troops." Ribbons were everywhere, and "thank you for your service" was on everyone's lips. And all but the military stayed home.

Today, as a retired member of the military, I am able to receive my health care at Walter Reed Hospital near Washington, D.C. Often when I am there for whatever aches and pains I may be having, I come across young soldiers in the hallways, sometimes walking or being pushed in a wheelchair, with their spouses and children. Some have no arms or no legs, or burn scars all over their bodies. I shrink in embarrassment for the relatively minor nature of my ailments compared with theirs, and am heartbroken for their young lives so terribly changed by war.

As I watched our responses in the years following the events of 9/11, I developed a sense of unease with what I felt was the often unconstrained development of new technologies for warfighting, and with a series of revelations that indicated senior administration officials' disregard for ethical behavior. Torture, kidnapping, and warrantless wiretapping didn't seem to square well with a nation that so vocally purported to value human rights. Guidance from national leaders to "take the gloves off" and "do whatever it takes" was reckless and ill-advised. We were acting like the American cowboys and gunslingers of a previous era, and the world noticed. National leaders played on the fear of the American people to justify enormous military expenditures. In his book *Mission Failure,* about U.S. policies after the 2001 attacks, Johns Hopkins University professor Michael Mandelbaum said, "If the government's imagination had failed it before September 11, afterward its imagination ran wild." Privacy and civil liberties, and at times possibly our decency, took a backseat to security and military force. I wondered on occasion whether we had lost our collective heart, if not our collective mind.

Fortunately, some of the hysteria and angst has abated over time, but with the recent surge in terror attacks around the globe and the growing militarism in Russia, China, and elsewhere, politicians are again trafficking in fear and voicing more urgent calls for increased military and intelligence capabilities. The key question is whether we can respond rationally, or if we will succumb to the urge to immediate action, brushing off the need for deliberations about future consequences. This begs the question, too, of whether safety and security are at all compatible with morality and ethics. I think they can be. Even in the face of all this complexity—new technologies, new enemies, societal changes, political gridlock—and a host of other challenges, it is possible to find a way to keep our moral compass and maintain our bearings.

If we delay much longer in debating and proscribing the use of radically new weapons, however, we may find there is no turning back once these systems get out of control. We're not talking about fantasy and science fiction here. We're talking about real problems generated by flawed or indecipherable software in billions of devices, of long-distance weapons that make violence too easy, of soldiers bred for combat, and of possibly unknown and uncontrollable pathogens. An arms race using all of the advanced technologies I've described will not be like anything we've seen, and the ethical implications are frightening.

Right now I do not believe the public has the necessary understanding of war or technology to be able to make sound judgments about either, or even to comprehend the potential future. We need—desperately—to change the status quo, with a wide-ranging national debate on the path forward. That debate cannot be left to the military alone. To do so is both unfair and unwise.

Our armed forces are critical to the safety and security of the country and its interests. We owe them much respect and gratitude. It is worth repeating, however, that rah-rah patriotism and facile gestures of support for our fighting men and women will no longer be enough. Because they are so important, the public can and must actively debate and decide what our military interests are and, critically, should know and influence when our forces fight and how they behave.

Despite the new threats facing us, the United States has an opportunity to show leadership in reducing tensions and somehow stemming a wasteful and dangerous arms race. Both of the inflection points, in technology and in our global military situation, provide us with a chance to reshape the future, or let it reshape us.

ACKNOWLEDGMENTS

This book is the product of several years of thinking, speaking, writing, and teaching on the subject of war and technology and the ethical dilemmas they create. I have not always been so interested or active in the field. Now that I have become concerned, I owe much to the active encouragement of my friends and colleagues at the University of Notre Dame's Reilly Center for Science, Technology, and Values in shaping my views. First among these is the late Jack Reilly, to whom this book is dedicated. Jack was excited to have these important topics publicly discussed. Other Reilly Center colleagues to whom I am indebted include directors Jerry McKenny and Anjan Chakravarty; my friend and former director Professor Don Howard, with whom I have collaborated, and who educated me in basic theories of philosophy; Jessica Baron, who worked tirelessly to raise public awareness; the graduate students, Matthew Lee, Charles Pence, and Pablo Ruiz, with whom I taught my course; and Kimberly Milewski, who is the corporate memory of the organization. I owe a special debt to Patrick McCloskey, with whom I serve on the Reilly advisory board and have published several articles. I hasten to add that the views expressed in this book are mine alone.

I also very much appreciate the support of Dr. Herb Lin, formerly at the National Academy of Sciences, and am extremely grateful to Samuel G. Freedman of Columbia University and *The New York Times,* whose article on me and my work generated much interest.

My daughter, Susan Latiff, a cognitive psychology Ph.D. student, always gave freely of her time to discuss the book with me and provide comments. And of course, I owe special thanks to Dale Latiff, my spouse of many years, who supported me and understood my sometimes total detachment during the extended process of developing and writing a book.

Finally, enormous thanks go to Jonathan Segal, my editor at Knopf, whose feedback taught me much about writing and whose patience with me was infinite.

NOTES

INTRODUCTION

4 In 1999, Chinese colonels: Qiao Liang and Wang Xiangsui, *Unrestricted Warfare* (Bejing: PLA Literature and Arts Publishing House, 1999), 121–31. The authors also have much to say about the seduction of militaries by technology.

1 THE NEW FACE OF WAR

18 Since 9/11, the military has experienced: Daniel Wirls, "Gridlock in Washington? Not When It Comes to Military Spending," *San Francisco Chronicle,* March 18, 2015. This article notes that $612 billion is 27 percent more than we spent in 2002 and equals our outlay at the height of the Cold War.

20 "many and simple": T. X. Hammes, "The Future of Warfare: Small, Many, Smart vs. Few and Exquisite?," *War on the Rocks,* July 16, 2014, https://warontherocks.com/2014/07/the-future-of-warfare-small-many-smart-vs-few-exquisite/. The author also has a lot to say about the relentlessly skyrocketing cost of "exquisite" weapons.

21 In 2014, the U.S. Army Research Laboratory: Alexander Kott et al., *Visualizing the Tactical Ground Battlefield in the Year 2050: Workshop Report* (Adelphi, MD: U.S. Army Research Laboratory, 2015).

21 The Defense Department is investing in technologies: Noah Shacht-

man and Robert Beckhusen, "11 Body Parts Defense Researchers Will Use to Track You," *Wired,* January 25, 2013.

22 The commander of the Navy's submarine forces: Megan Eckstein, "COMSUBFOR Connor: Submarine Force Could Become the New A2/AD Threat," *U.S. Naval Institute News,* May 14, 2015.

23 The Defense Department's inventory: Jeremiah Gertler, *U.S. Unmanned Aerial Systems* (Washington, DC: Congressional Research Service, 2012). While this document only discusses aerial systems, other services are investing heavily in unmanned systems as well.

23 Video surveillance was a $2 billion industry: Organisation for Economic Co-operation and Development, *The Security Economy* (Paris: OECD Publications, 2004).

23 It reached $21 billion: Transparency Market Research, *Video Surveillance and VSaaS Market—Global Industry Analysis, Size, Share, Growth, Trends and Forecast, 2016–2024,* April 29, 2016, http://www.transparencymarketresearch.com/video-surveillance-vsaas-market.html.

24 War will not necessarily be fought: Thomas Gibbons-Neff, "The New Type of War That Finally Has the Pentagon's Attention," *Washington Post,* July 3, 2015.

25 For the future, the Department of Defense lists: Robin Staffin, "Department of Defense Basic Research," Presentation to National Defense Industrial Association, June 19, 2013.

25 In addition, the DOD plans: Earl Wyatt, "Prototyping: A Path to Agility, Innovation, and Affordability," Presentation to National Defense Industrial Association, March 24, 2015.

26 As the columnist David Ignatius summarizes: David Ignatius, "Arming Ourselves for the Next War," *Washington Post,* February 24, 2016.

28 The amount of data now gathered: M. G. Siegler and Eric Schmidt, "Every 2 Days We Create as Much Information as We Did Up to 2003," *Tech Crunch,* August 4, 2010.

28 Appliances, vehicles, and even toys: Mark P. Mills, "Creepy Barbie? Brace Yourself for the Internet of Toys," *Forbes,* December 22, 2015.

This article also discusses the many implications of data collected without the consumer's explicit knowledge.

28 For several years, the Army issued: Jon Hamilton, "Pentagon Shelves Blast Gauges Meant to Detect Battlefield Brain Injuries," NPR, December 20, 2016.

29 Such luminaries as Bill Gates: Cecilla Tilli, "Killer Robots? Lost Jobs? The Threats That Artificial Intelligence Researchers Actually Worry About," *Slate,* April 28, 2016.

29 A former chief technology officer: Michael B. Kelley, "CIA Chief Tech Officer: Big Data Is the Future and We Own It," *Business Insider,* March 21, 2013.

30 History has demonstrated that in times of crisis: James Waldo, Herbert Lin, and Lynette I. Millett, *Engaging Privacy and Information Technology in a Digital Age* (Washington, DC: National Academies Press, 2007), 349–65.

30 Synthetic biology is an emerging area: Royal Society, "Call for Views: Synthetic Biology," June 2007, https://royalsociety.org/~/media/Royal_Society_Content/policy/projects/synthetic-biology/CallForViews.pdf.

30 It holds great promise: Office of Technical Intelligence, *Technical Assessment: Synthetic Biology* (Washington, DC: Department of Defense, January 2015).

32 So worrisome is the CRISPR capability: Kathryn Ziden, "The Dark Side of CRISPR," Potomac Institute for Policy Studies Center for Revolutionary Scientific Thought, September 20, 2016.

32 In an interview with Stanford University writer: Mark Shwartz, "Biological Warfare Emerges as 21st-Century Threat," *Stanford Report,* January 11, 2001.

32 Some scientists believe it is possible: See Douglas R. Lewis, "An Era of Hopes and Fears," *Strategic Studies Quarterly* 10, no. 3 (2016): 23–46.

32 Indeed, in a widely reported case: John P. Geiss II and Theodore C. Hailes, "Deterring Emergent Technologies," *Strategic Studies Quarterly* 10, no. 3 (2016): 47–73.

34 Of further concern: Michael Specter, "A Life of Its Own: Where Will Synthetic Biology Lead Us?," *The New Yorker,* September 28, 2009.

35 If brain function is subsequently destroyed: Robbin A. Miranda et al., "DARPA-Funded Efforts in the Development of Novel Brain-Computer Interface Technologies," *Journal of Neuroscience Methods* 244 (2015): 52–67.

35 Other researchers have now shown: Susan Young Rojhan, "Rats Communicate Through Brain Chips," *MIT Technology Review,* February 28, 2013.

37 It would certainly give new meaning: Andrea Stocco et al., "Playing 20 Questions with the Mind: Collaborative Problem Solving by Humans Using a Brain-to-Brain Interface," *PloS One* 10, no. 9 (2015): e0137303.

41 The deputy secretary of defense, Robert Work: Sydney J. Freedberg Jr., "Will US Pursue 'Enhanced Human Ops?' DepSecDef Wonders," *Breaking Defense,* December 14, 2015.

42 The bioethicist Jonathan Moreno: Jonathan Moreno, "DARPA on Your Mind," *Neuroethics Publications* (2004): 30.

42 Author and former Marine: Karl Marlantes, *What It Is Like to Go to War* (New York: Atlantic Monthly Press, 2011), 232.

46 Notwithstanding the stated policy: Thomas K. Adams, "Future Warfare and the Decline of Human Decisionmaking," *Parameters* 31, no. 4 (2001): 57.

47 In 2006, the U.S. Army surgeon general: Department of the Army, *Mental Health Advisory Team (MHAT) IV: Operation Iraqi Freedom 05-07,* U.S. Army Surgeon General's Office, November 16, 2006.

47 Georgia Tech University professor: Ronald C. Arkin, *Governing Lethal Behavior: Embedding Ethics in a Hybrid Deliberative/Reactive Robot Architecture* (Atlanta: Georgia Tech University, 2007), http://www.cc.gatech.edu/ai/robot-lab/onlinepublications/formalizationv35.pdf.

47 There is even a global effort: Stephen Goose, "The Case for Banning Killer Robots," Human Rights Watch, November 24, 2015, https://www.hrw.org/news/2015/11/24/case-banning-killer-robots.

47 Underlying many of the concerns: Kenneth Anderson and Matthew Waxman, "Law and Ethics for Autonomous Weapon Systems: Why a Ban Won't Work and How the Laws of War Can," Hoover Institution, Stanford University, 2013, https://ssrn.com/abstract=2250126.

48 The Department of Defense says: Department of Defense Directive 3000.09, Autonomy in Weapon Systems, November 21, 2012.

48 Pentagon planners are quoted: Eric Beidel, Sandra I. Erwin, and Stew Magnuson, "10 Technologies the U.S. Military Will Need for the Next War," *National Defense,* November 2011.

48 In 1988, the U.S. guided missile cruiser: "Iran Air Flight 655," World eBook Library, http://www.ebooklibrary.org/articles/Iran_Air_Flight_655. See also Nancy C. Roberts, *Reconstructing Combat Decisions: Reflections on the Shootdown of Flight 655* (Monterey, CA: Naval Postgraduate School, October 1992).

49 Air Force chief scientist Greg Zacharias: Phillip Swarts, "Air Force Looking at Autonomous Systems to Aid War Fighters," *Air Force Times,* May 17, 2016.

49 In an even more technically aggressive project: John Keller, "DARPA Rounds Out Gremlins Program with Four Companies to Create Overwhelming Drone Swarms," *Military and Aerospace Electronics,* May 10, 2016.

50 This is not as far-fetched: Robert D. Mulcahy Jr., ed., *Corona Star Catchers* (Washington, DC: Center for the Study of National Reconnaissance, June 2012), http://www.nro.gov/history/csnr/corona/StarCatchersWeb.pdf.

50 These swarms, they say: See Paul Scharre and Shawn Brimley, "20YY: The Future of Warfare," *War on the Rocks,* January 29, 2014.

50 Rebels, in addition to: See Bryan Bender, "The Secret U.S. Army Study That Targets Moscow," *Politico,* April 14, 2016.

51 Army major Daniel Sukman: Daniel Sukman, "Lethal Autonomous Systems and the Future of Warfare," *Canadian Military Journal* 16, no. 1 (Winter 2015): 44–53.

52 There have been instances: Kim Zetter, "Feds Say That Banned Re-

searcher Commandeered a Plane," *Wired,* May 15, 2015. Joseph Bergermarch, "A Dam, Small and Unsung, Is Caught Up in an Iranian Hacking Case," *The New York Times,* March 25, 2016.

54 The Department of Defense has decreed: Patrick Tucker, "NSA Chief: Rules of War Apply to Cyberwar Too," *Defense One,* April 20, 2015.

54 However, the chief of the U.S. Army's Cyber Center of Excellence: Sandra Jontz, "Cyber Ethics Vex Online Warfighters," *Signal,* January 1, 2016.

55 The U.S. Air Force designed and built: George I. Seffers, "CHAMP Prepares for Future Fights," *Signal,* February 1, 2016.

61 Yet rapid technology growth itself poses: Stewart Brand, "Is Technology Moving Too Fast?," *Time,* June 19, 2000.

62 If the situation of dizzying complexity: Samuel Arbesman, "It's Complicated," *Aeon,* January 6, 2014.

2 HOW WE GOT TO NOW

66 Happily, the total numbers of dead: Colin Schulz, "Globally, Deaths from War and Murder Are in Decline," *Smithsonian,* March 21, 2014.

68 As defined by the U.S. Special Operations Command: Philip Kapusta, "Gray Zone," *Special Warfare* 28, no. 4 (October–December 2015): 18–25.

70 Some of the first presentations: Brenda J. Buchanan, ed., "Editor's Introduction," in *Gunpowder, Explosives, and the State: A Technological History* (New York: Routledge, 2006).

70 The development of weapons follows: "Punctuated Equilibrium," Evolution Library, PBS, http://www.pbs.org/wgbh/evolution/library/03/5/l_035_01.html.

73 The civil space program: Mulcahy, *Corona Star Catchers.*

73 Military command and control systems depend: Rick Lober, "Why the Military Needs Commercial Satellite Technology," *Defense One,* September 25, 2013.

75 In his book *Technopoly:* Nancy Kaplan, "Postman," *Computer-Mediated Communication* 2, no. 3 (March 1, 1995): 23.

75 The physicist Robert Oppenheimer: James A. Hijiya, "The Gita of J. Robert Oppenheimer," *Proceedings of the American Philosophical Society* 144, no. 2 (June 2000): 123–67.

77 One of the U.S. military's top combat commanders: John Dickerson, "A Marine General at War," *Slate,* April 22, 2010.

78 The technology and culture writer: Tara Parker-Pope, "An Ugly Toll of Technology: Impatience and Forgetfulness," *The New York Times,* June 6, 2010.

78 For example, many consider ballistic missile defense: Joseph Cirincione, "Brief History of Ballistic Missile Defense and Current Programs in the United States," Testimony, Carnegie Endowment for International Peace, February 1, 2000, http://carnegieendowment.org/2000/01/31/brief-history-of-ballistic-missile-defense-and-current-programs-in-united-states-pub-133.

78 The random failure of a communications satellite: Laurence Zuckerman, "Satellite Failure Is Rare, and Therefore Unsettling," *The New York Times,* May 21, 1998.

78 The loss of GPS satellite navigation: "Friendly Fire Kills Three US Soldiers," *Guardian,* December 5, 2001.

80 Antibiotics are miracle drugs: Michael Enright, "The Looming Crisis of Antibiotic Resistance," *The Sunday Edition,* CBC Radio, August 30, 2015.

81 Chemicals enhance the food supply: Jacque Wilson and Jen Christensen, "7 Other Chemicals in Your Food," CNN, February 10, 2014.

81 British neuroscientist and policy adviser: Susan Greenfield, "Modern Technology Is Changing the Way We Think," *Daily Mail,* December 30, 2015.

82 The United States spends astronomical sums: Lauren Carroll, "Obama: US Spends More on Military Than Next 8 Nations Combined," *Politifact,* January 13, 2016.

84 The LRSO seems to be: Hams M. Kristensen, "LRSO: The Nuclear

Cruise Missile Mission," Federation of American Scientists, October 20, 2015.

84 Worse yet are cases of shoddy workmanship: Howard Zinn, "Robber Barons and Rebels," chapter 11 in *History Is a Weapon: A People's History of the United States* (New York: HarperCollins, 1980).

84 More recently, the Defense Department inspector general: James Risen, "Despite Alert, Flawed Wiring Still Kills G.I.'s," *The New York Times,* May 4, 2008.

84 A 2014 study by Bloomberg: Richard Clough, "U.S. Defense Industry's Profits Soaring Along with Global Tensions," Bloomberg News, September 25, 2014.

84 The project is reportedly responsible: Jeremy Bender, Armin Rosen, and Skye Gould, "This Map Shows Why the F-35 Has Turned Into a Trillion-Dollar Fiasco," *Business Insider,* August 20, 2014.

85 To date, the fighter is still short: Directorate of Operational Test and Evaluation, "Joint Strike Fighter (JSF)," *FY 2015 DOD Programs* (Washington, DC: Department of Defense, 2015), http://www.dote.osd.mil/pub/reports/FY2015/pdf/dod/2015f35jsf.pdf.

85 Essayist James Fallows: James Fallows, "The Tragedy of the American Military," *The Atlantic,* January–February 2015.

86 We are huge weapons proliferators: "International Arms Transfers," Stockholm International Peace Research Institute, https://www.sipri.org/research/armament-and-disarmament/arms-transfers-and-military-spending/international-arms-transfers.

86 In 2012, then national security adviser: John O. Brennan, "The Ethics and Efficacy of the President's Counterterrorism Strategy," remarks at the Wilson Center, April 30, 2012, https://www.wilsoncenter.org/event/the-efficacy-and-ethics-us-counterterrorism-strategy.

87 It is widely believed: Bart Jansen, "Report Confirms MH17 Shot Down—But Why?," *USA Today,* October 14, 2015.

88 "If we get to a point": David Sterman, "Will We Still Call It War?," *Time,* March 9, 2015.

3 EFFECTS OF FUTURE WAR ON THE SOLDIER

93 Author and activist: Peter Marin, "Living in Moral Pain," *Psychology Today,* November 1981.

93 General John Allen: George R. Lucas, *Military Ethics: What Everyone Needs to Know* (New York: Oxford University Press, 2016), 17.

93 The Just War philosopher: Michael Walzer, *Just and Unjust Wars: A Moral Argument with Historical Illustrations* (New York: Basic Books, 2015).

94 In a discussion of the distinction: Maggie Puniewska, "Healing a Wounded Sense of Morality," *The Atlantic,* July 3, 2015.

95 Images and descriptions: Nicholas D. Kristof, "Unmasking Horror—A Special Report.; Japan Confronting Gruesome War Atrocity," *The New York Times,* March 17, 1995.

96 Australian ethicist: Robert Sparrow, "War Without Virtue?" in *Killing by Remote Control: The Ethics of an Unmanned Military,* ed. Bradley Jay Strawser (New York: Oxford University Press, 2013), 84–105.

97 Notre Dame professor Don Howard: Robert H. Latiff and Don Howard, *Ethical, Legal, and Societal Implications of New Weapons Technologies: A Briefing Book for Presenters* (Washington, DC: National Academy of Sciences, 2015).

97 The Dutch statesman and scholar: Hugo Grotius, *The Rights of War and Peace, Including the Law of Nature and of Nations, Translated from the Original Latin of Grotius, with Notes and Illustrations from Political and Legal Writers, by A. C. Campbell, A.M., with an Introduction by David J. Hill* (New York: M. Walter Dunne, 1901), accessed online on May 17, 2016, http://oll.libertyfund.org/titles/553.

98 The rules of war first found expression: Latiff and Howard, *Ethical, Legal, and Societal Implications of New Weapons Technologies.*

98 Lincoln's War Department was deeply concerned: Paul Finkelman, "Francis Lieber and the Modern Law of War" (reviewing John Fabian Witt, *Lincoln's Code: The Laws of War in American History*), *University of Chicago Law Review* 80, no. 4 (September 2013): 2071–132.

99 Scientists attending the Asilomar Conference: Paul Berg, "Meetings That Changed the World: Asilomar 1975: DNA Modification Secured," *Nature* 455 (September 2008): 290–91.

101 The medieval code of chivalry included: Richard Abels, "Medieval Chivalry," United States Naval Academy, http://www.usna.edu/Users/history/abels/hh315/Chivalry.htm.

101 The Dutch historian: Johan Huizinga, *The Waning of the Middle Ages* (Mineola, NY: Dover Publications, 1999; originally published in 1919), 56–65.

101 Oxford University law professor: Theodor Meron, "International Humanitarian Law from Agincourt to Rome," *International Law Studies Across the Spectrum of Conflict* 75 (2000), https://www.usnwc.edu/Research---Gaming/International-Law/New-International-Law-Studies-(Blue-Book)-Series/International-Law-Blue-Book-Articles.aspx?Volume=75.

101 These may seem like quaint notions: Theodor Meron, *Bloody Constraint: War and Chivalry in Shakespeare* (New York: Oxford University Press, 1998), 118.

102 The case of Lieutenant Calley: Seymour Hersh, "Lieutenant Accused of Murdering 109 Civilians," *St. Louis Post-Dispatch,* November 13, 1969. See also Tom Ricks, *The Generals: American Military Command from World War II to Today* (New York: Penguin, 2012).

102 World War II general Curtis LeMay: Michael Sherry, *The Rise of American Air Power: The Creation of Armageddon* (New Haven, CT: Yale University Press, 1989), 287.

103 As Iraqi forces were on the run: Seymour Hersh, "Overwhelming Force: What Happened in the Final Days of the Gulf War?," *The New Yorker,* May 15, 2000.

106 In the face of the brutality: Sara Wood, "Gen. Petraeus Urges Troops to Adhere to Ethical Standards," American Forces Press Service, May 14, 2007.

107 They feel that ethics in war: Ralph Peters, "A Revolution in Military Ethics?," *Parameters* 26, no. 2 (1996): 102.

107 The philosopher Michael Walzer: Walzer, *Just and Unjust Wars,* 45.

107 The Canadian author and former politician: Michael Ignatieff, "Reimagining a Global Ethic," *Ethics and International Affairs,* Carnegie Council, April 1, 2012.

112 Former naval intelligence officer William Bray: William R. Bray, "Man Versus Machine," *Signal,* December 1, 2016.

113 The philosopher Simone Weil: Mary McCarthy and Simone Weil, "The Iliad, or the Poem of Force," *Chicago Review* 18, no. 2 (1965): 5–30.

114 Human perception and coordination: Adams, "Future Warfare and the Decline of Human Decisionmaking."

115 The philosopher Arnold Toynbee: Arnold Toynbee, "Why I Dislike Western Civilization," *The New York Times,* May 10, 1964.

117 "A robot's targets": Chris Baraniuk, "World War R: Rise of the Killer Robots," *New Scientist,* November 15, 2014.

117 Death by algorithm: Robert H. Latiff and Patrick J. McCloskey, "With Drone Warfare, America Approaches the Robo-Rubicon," *The Wall Street Journal,* March 14, 2013.

117 Echoing the concerns of senior combat leaders: Janine Davidson, "The Warrior Ethos at Risk: H. R. McMaster's Remarkable Veterans Day Speech," *Defense in Depth* blog, Council on Foreign Relations, November 18, 2014, http://blogs.cfr.org/davidson/2014/11/18/the-warrior-ethos-at-risk-h-r-mcmasters-remarkable-veterans-day-speech/.

118 Yale University ethicist: Wendell Wallach, *A Dangerous Master: How to Keep Technology from Slipping Beyond Our Control* (New York: Basic Books, 2014).

119 DARPA has recently begun to address: Jean-Lou Chameau, William F. Ballhaus, and Herbert S. Lin, *Emerging and Readily Available Technologies and National Security: A Framework for Addressing Ethical, Legal, and Societal Issues* (Washington, DC: National Academies Press, 2014).

120 Another member of the committee: Baruch Fischhoff, "Ethical and Social Issues in Military Research and Development," *Telos* 169 (2014): 150–54.

4 SOCIETY AND THE MILITARY

123 We saw access to the former Soviet states: Interview with Thomas Graham, *Frontline,* PBS, http://www.pbs.org/wgbh/pages/frontline/shows/yeltsin/interviews/graham.html.

124 Wendell Wallach noted that: Wallach, *A Dangerous Master,* 60.

124 British author Christopher Coker: Christopher Coker, *Ethics and War in the 21st Century* (London: Routledge, 2008), 156.

125 Many of the more strident supporters: MacGregor Knox and Williamson Murray, eds. *The Dynamics of Military Revolution, 1300–2050* (New York: Cambridge University Press, 2001), 190.

126 At a 1993 meeting: Walter Isaacson, "Madeleine's War," *Time,* May 17, 1999.

127 "There was a vision": James Mann, *Rise of the Vulcans: The History of Bush's War Cabinet* (New York: Penguin, 2004), xii.

127 In writing about how the United States should respond: Charles Krauthammer, "How Fast Things Change," *Townhall,* November 30, 2001, https://townhall.com/columnists/charleskrauthammer/2001/11/30/how-fast-things-change-n1401085.

127 In his book *Virtual War:* Michael Ignatieff, *Virtual War: Kosovo and Beyond* (New York: Picador, 2000), 215.

128 British academic Sir Alistair Horne: Alistair Horne, *Hubris: The Tragedy of War in the Twentieth Century* (New York: HarperCollins, 2015).

131 Russian media official Dmitry Kiselyov: Derek Bacon, "Yes, I'd Lie to You," *The Economist,* September 10, 2016.

132 As former military intelligence officer Jim Gourley describes: Jim Gourley, "Welcome to Spartanburg!: The Dangers of This Growing American Military Obsession," *Foreign Policy,* April 22, 2014.

133 Given the well-documented conservative leanings: Frank Newport, "Military Veterans of All Ages Tend to Be More Republican," Gallup, May 25, 2009, http://www.gallup.com/poll/118684/military-veterans-ages-tend-republican.aspx.

133 As Marine pilot: Carl Forsling, "If You Call All Veterans Heroes, You're Getting It Wrong," *Task and Purpose,* August 5, 2014.

133 In the words of retired Air Force: William J. Astore, "Every Soldier a Hero? Hardly," *Los Angeles Times,* July 22, 2010.

134 The United States now ranks twelfth: Ray Williams, "The Cult of Ignorance in the U.S.: Anti-Intellectualism and the 'Dumbing Down' of America," *Progreso Weekly,* May 29, 2016, http://progresoweekly.us/cult-ignorance-u-s-anti-intellectualism-dumbing-america/.

135 With health care costs consuming: Olga Khazan, "U.S. Healthcare: Most Expensive and Worst Performing," *The Atlantic,* June 16, 2014.

135 In 2009, only eleven states: Russell Berman, "The Real Language Crisis," American Association of University Professors, September–October 2011, https://www.aaup.org/article/real-language-crisis#.WKtcexB4O8Y.

136 In Thucydides' *History of the Peloponnesian War:* Thucydides, *History of the Peloponnesian War,* translated by Rex Warner (New York: Penguin, 1972), 213.

136 International security scholar: J. Peter Scoblic, "Presidents Need to Be Able to Do Nothing," *Washington Post,* July 15, 2016.

136 William Manchester, in his book: William Manchester, *A World Lit Only by Fire: The Medieval Mind and the Renaissance; Portrait of an Age* (New York: Back Bay Books, 1992).

137 In the current Internet age: Williams, "The Cult of Ignorance in the U.S."

137 His critics believed that intellect: Susan Searls Giroux, "Between Race and Reason: Anti-Intellectualism in American Life," *Truthout,* September 16, 2011.

137 In the 1980s, President Reagan: Andrew J. Bacevich, *The New American Militarism: How Americans Are Seduced by War* (New York: Oxford University Press, 2013), 103.

139 In the seventy years: Barbara Salazar Torreon, *Instances of Use of United States Armed Forces Abroad, 1798–2016* (Washington, DC: Congressional Research Service, October 2016).

139 The number of elected officials: Jennifer Rizzo, "Veterans in Congress at Lowest Level Since World War II," CNN, January 21, 2011, http://www.cnn.com/2011/POLITICS/01/20/congress.veterans/.

139 Just one-third of veterans: "War and Sacrifice in the Post-9/11 Era," Pew Research Center, October 5, 2011, http://www.pewsocialtrends.org/2011/10/05/war-and-sacrifice-in-the-post-911-era/.

139 Iraq combat veteran: Patrick J. Murphy, "A Soldier Reflects: Those Who Had the Least to Lose Sent Us to War," MSNBC, March 16, 2013.

139 "What was the point?" one asked: Bill Briggs, " 'I Risked My Life, for What?': Iraq War Veterans Chilled by Country's Slide into Civil War," NBC News, July 25, 2013.

139 In 2011, the Pew Research Center: "War and Sacrifice in the Post-9/11 Era."

141 In a recent survey: Asma Khalid, "Millennials Want to Send Troops to Fight ISIS, but Don't Want to Serve," NPR, December 10, 2015, http://www.npr.org/2015/12/10/459111960/millennials-want-to-send-troops-to-fight-isis-but-not-serve.

142 Minorities represent a larger percentage: Jeremy Bender, "These 22 Charts Reveal Who Serves in America's Military," *Business Insider*, August 14, 2014.

142 Retired Army general Karl Eikenberry: Karl W. Eikenberry and David M. Kennedy, "Americans and Their Military, Drifting Apart," *The New York Times*, May 26, 2013.

142 Retired admiral Mike Mullen: Fallows, "The Tragedy of the American Military."

143 In a 2003 poll of military members: Gourley, "Welcome to Spartanburg!"

5 WHERE DO WE GO FROM HERE?

146 The situation is serious: See, for example, Rosa Brooks, *How Everything Became War and the Military Became Everything* (New York: Simon & Schuster, 2016); Rachel Maddow, *Drift: The Unmooring of American Military Power* (New York: Broadway Books, 2012); and

Andrew J. Bacevich, *The New American Militarism: How Americans Are Seduced by War* (New York: Oxford University Press, 2013).

146 "The president has been commander in chief": Robert McMahon, "Balance of War Powers: The U.S. President and Congress," *CFR Backgrounders,* Council on Foreign Relations, June 20, 2011, http://www.cfr.org/united-states/balance-war-powers-us-president-congress/p13092.

147 Conor Friedersdorf, writing in *The Atlantic:* Conor Friedersdorf, "The Congress Shall Have the Power . . . to Declare War," *The Atlantic,* August 27, 2014.

148 In 2009, lawmakers tried to use OCO funds: Emerson Brooking and Janine Davidson, "How the Overseas Contingency Operations Fund Works—and Why Congress Wants to Make It Bigger," *Defense in Depth* blog, Council on Foreign Relations, June 16, 2015, http://blogs.cfr.org/davidson/2015/06/16/how-the-overseas-contingency-operations-fund-works-and-why-congress-wants-to-make-it-bigger/.

149 The Government Accountability Office: Chris Edwards and Nicole Kaeding, "Federal Government Cost Overruns," *Tax and Budget Bulletin* 72 (Washington, DC: Cato Institute, September 2015).

150 Another GAO study reported: Sandra I. Irwin, "Weapon Cost Overruns: From Bad to Worse," *National Defense,* January 2009.

151 In an amazing series of articles: Dana Priest and William Arkin, "Top Secret America," *The Washington Post,* July 20, 2010.

151 The 561,000 service contractors: David Lerman, "Fiscal Follies," *CQ Budget News,* October 6, 2016.

152 Admiral Mike Mullen has said: Fallows, "The Tragedy of the American Military."

156 Expanding further on the idea: Jason Mangone, "National Service Is the Proper Response to National Emergency," *Huffington Post,* September 11, 2014.

156 Far more impactful would be: Eric Navarro, "Could One Year of Mandatory National Service Change This Country?," *Task and Purpose,* June 9, 2014.

A CONCLUDING PLEA

161 In his book *Mission Failure:* Michael Mandelbaum, *Mission Failure: America and the World in the Post–Cold War Era* (New York: Oxford University Press, 2016), 144.

INDEX

A NOTE ABOUT THE AUTHOR

Robert H. Latiff is a retired U.S. Air Force major general. He is a private consultant to corporations, universities, and government agencies. He is associated with the University of Notre Dame's Reilly Center for Science, Technology, and Values as chairman of its advisory board and as an adjunct professor. He has commanded at multiple levels in the U.S. Army and the U.S. Air Force, and has served on the staff of the secretary of the Air Force. He is a member of the Air Force Studies Board and the Intelligence Community Studies Board of the National Academy of Sciences. He has contributed chapters for edited volumes on military ethics, and has written editorials for *The Wall Street Journal,* Fox News, and CNN. General Latiff holds a Ph.D. in engineering from the University of Notre Dame. He resides in Alexandria, Virginia.

A NOTE ON THE TYPE

This book was set in Adobe Garamond. Designed for the Adobe Corporation by Robert Slimbach, the fonts are based on types first cut by Claude Garamond (c. 1480–1561). Garamond was a pupil of Geoffroy Tory and is believed to have followed the Venetian models, although he introduced a number of important differences, and it is to him that we owe the letter we now know as "old style." He gave to his letters a certain elegance and feeling of movement that won their creator an immediate reputation and the patronage of Francis I of France.

Typeset by Scribe,
Philadelphia, Pennsylvania

Printed and bound by
LSC / Harrisonburg North,
Harrisonburg, Virginia

Designed by Maggie Hinders